The
Economist

NUMBERS GUIDE

OTHER ECONOMIST BOOKS

Guide to Analysing Companies
Guide to Business Modelling
Guide to Economic Indicators
Guide to Financial Markets
Guide to the European Union
Guide to Management Ideas
Style Guide
Dictionary of Business
Dictionary of Economics
International Dictionary of Finance
Business Ethics
China's Stockmarket
E-Commerce
E-Trends
Economics
Globalisation
Measuring Business Performance
Successful Innovation
Successful Mergers
Wall Street

Pocket Asia
Pocket Europe in Figures
Pocket World in Figures

The Economist

NUMBERS GUIDE

The Essentials of Business Numeracy

THE ECONOMIST IN ASSOCIATION WITH
PROFILE BOOKS LTD

Published by Profile Books Ltd
3A Exmouth House, Pine Street, London ECIR OJH

First published by The Economist Books Ltd 1991

Typeset by International Typesetters Inc.
info@InternationalTypesetters.com

Printed in Great Britain by
St Edmundsbury Press, Bury St Edmunds

A CIP catalogue record for this book is available
from the British Library

ISBN-13: 978-1 86197 515-7
ISBN-10: 1 86197 515-5

For information on other Economist Books, visit
www.profilebooks.co.uk
www.economist.com

Contents

List of tables

List of figures

Introduction

*"Statistical thinking will one day be as necessary a
qualification for efficient citizenship as
the ability to read and write."*

H.G. Wells

THIS BOOK is about solving problems and making decisions using
numerical methods. Everyone – people in business, social adminis-
trators, bankers – can do their jobs better if equipped with such tools. No
special skills or prior knowledge are required. Numerical methods
amount to little more than applied logic: they all reduce to step-by-step
instructions and can be processed by simple computing devices. Yet
numerical methods are exciting and powerful. They work magic, which
is perhaps why they are shrouded in mystery. This book strips away
that mystery and provides a guided tour through the statistical workshop. There are no secrets, no barriers to entry. Anyone can use these
tools. Everyone should.

What are numerical methods?

Numerical methods range from the simple (how to calculate percentages and interest) to the relatively complex (how to evaluate competing
investment opportunities); from the concrete (how to find the shortest
route for deliveries) to the vague (how to deal with possible levels of
sales or market share). The link is quantitative analysis, a scientific
approach.

This does not mean that qualitative factors (intangibles such as personal opinion, hunch, technological change and environmental aware-
ness) should be ignored. On the contrary, they must be brought into the
decision process, but in a clear, unemotional way. Thus, a major part of
this book is devoted to dealing with risk. After all, this forms a major
part of the business environment. Quantifying risk and incorporating it
into the decision-making process is essential for successful business.

In bringing together quantitative techniques, the book borrows heavily from mathematics and statistics and also from other fields, such as
accounting and economics.

A brief summary

We all perform better when we understand why we are doing something. For this reason, this book always attempts to explain why as well as how methods work. Proofs are not given in the rigorous mathematical sense, but the techniques are explained in such a way that the reader should be able to gain at least an intuitive understanding of why they work. This should also aid students who use this book as an introduction to heavier statistical or mathematical works.

The techniques are illustrated with business examples where possible but sometimes abstract illustrations are preferred. This is particularly true of probability, which is useful for assessing business risk but easier to understand through gamblers' playing cards and coins.

Examples use many different currencies and both metric and imperial measurements. The si standards for measurement (see SI UNITS) in the A–Z are excellent, but they are generally ignored here in favour of notation and units which are more familiar.

This book works from the general to the particular.

Chapter 1 lays the groundwork by running over some key concepts. Items of particular interest include proportions and percentages (which appear in many problems) and probability (which forms a basis for assessing risk).

Chapter 2 examines ways of dealing with problems and decisions involving money, as many or most do. Interest, inflation and exchange rates are all covered. Note that the proportions met in the previous chapter are used as a basis for calculating interest and evaluating investment projects.

Chapter 3 looks at summary measures (such as averages) which are important tools for interpretation and analysis. In particular, they unlock what is called the normal distribution, which is invaluable for modelling risk.

Chapter 4 reviews the way data are ordered and interpreted using charts and tables. A series of illustrations draws attention to the benefits and shortfalls of various types of presentation.

Chapter 5 examines the vast topic of forecasting. Few jobs can be done successfully without peering into the future. The objective is to pull together a view of the future in order to enhance the inputs to decision-making.

Chapter 6 marks a turning point. It starts by considering the way that sampling saves time and money when collecting the inputs to decisions. This is a continuation of the theme in the previous chapters. However, the chapter then goes on to look at ways of reaching the best decision from sample data. The techniques are important for better decision-making in general.

Chapter 7 expands on the decision theme. It combines judgment with the rigour of numerical methods for better decisions in those cases which involve uncertainty and risk.

Chapter 8 looks at some rather exciting applications of techniques already discussed. It covers:

- game strategy (for decision-making in competitive situations);
- queueing (for dealing with a wide range of business problems, only one of which involves customers waiting in line);
- stock control (critical for minimising costs);
- Markov chains (for handling situations where events in the future are directly affected by preceding events);
- project management (with particular attention to risk); and
- simulation (for trying out business ideas without risking humiliation or loss).

Chapter 9 reviews powerful methods for reaching the best possible decision when risk is not a key factor.

An A-z section concludes the book. It gives key definitions. It covers a few terms which do not have a place in the main body of the book. And it provides some useful reference material, such as conversion factors and formulae for calculating areas and volumes.

Additional information is available on this book's website at www.NumbersGuide.com.

How to use this book

There are four main approaches to using this book.

1 If you want to know the meaning of a mathematical or statistical term, consult the A-z. If you want further information, turn to the page referenced in the A-z entry and shown in small capital letters, and read more.
2 If you want to know about a particular numerical method, turn to the appropriate chapter and read all about it.
3 If you have a business problem that needs solving, use the A-z, the contents page, or this chapter for guidance on the methods available, then delve deeper.

4 If you are familiar with what to do but have forgotten the detail, then formulae and other reference material are highlighted throughout the book.

Calculators and PCs

There can be few people who are not familiar with electronic calculators. If you are selecting a new calculator, choose one with basic operations (+ − × and ÷) and at least one memory, together with the following.

- Exponents and roots (probably summoned by keys marked x^y and $x^{1/y}$): essential for dealing with growth rates, compound interest and inflation.
- Factorials (look for a key such as x!): useful for calculating permutations and combinations.
- Logarithms (log and 10^x or ln and e^x): less important but sometimes useful.
- Trigonometric functions (sin, cos and tan): again, not essential, but handy for some calculations (see TRIANGLES AND TRIGONOMETRY).
- Constants π and e: occasionally useful.
- Net present value and internal rate of return (NPV and IRR). These are found only on financial calculators. They assist investment evaluation, but you will probably prefer to use a spreadsheet.

PC users will find themselves turning to a spreadsheet program to try out many of the techniques in this book. PC spreadsheets take the tedium out of many operations and are more or less essential for some activities such as simulation.

For non-PC users, a spreadsheet is like a huge sheet of blank paper divided up into little boxes (known as cells). You can key text, numbers or instructions into any of the cells. If you enter ten different values in a sequence of ten cells, you can then enter an instruction in the eleventh, perhaps telling the PC to add or multiply the ten values together. One powerful feature is that you can copy things from one cell to another almost effortlessly. Tedious, repetitive tasks become simple. Another handy feature is the large selection of instructions (or functions) which enable you to do much more complex things than you would with a calculator. Lastly, spreadsheets also produce charts which are handy for interpretation and review.

The market-leader in spreadsheet programs is Microsoft Excel (packaged with Microsoft Office). It is on the majority of corporate desktops. However, if you are thinking about buying, it is worth looking at Sun Microsystems' Star Office and IBM's Lotus SmartSuite. Both of these options are claimed to be fully compatible with Microsoft Office.

Conclusion

There are so many numerical methods and potential business problems that it is impossible to cross-reference them all. Use this book to locate techniques applicable to your problems and take the following steps:

- Define the problem clearly.
- Identify the appropriate technique.
- Collect the necessary data.
- Develop a solution.
- Analyse the results.
- Start again if necessary or implement the results.

The development of sophisticated computer packages has made it easy for anyone to run regressions, to identify relationships or to make forecasts. But averages and trends often conceal more than they reveal. Never accept results out of hand. Always question whether your analysis may have led you to a faulty solution. For example, you may be correct in noting a link between national alcohol consumption and business failures; but is one directly related to the other, or are they both linked to some unidentified third factor?

1 Key concepts

"Round numbers are always false."

Samuel Johnson

Summary
Handling numbers is not difficult. However, it is important to be clear about the basics. Get them right and everything else slots neatly into place.

People tend to be comfortable with percentages, but it is very easy to perform many calculations using proportions. The two are used interchangeably throughout this book. When a result is obtained as a proportion, such as 6 ÷ 100 = 0.06, this is often referred to as 6%. Sums become much easier when you can convert between percentages and proportions by eye: just shift the decimal point two places along the line (adding zeros if necessary).

Proportions simplify problems involving growth, reflecting perhaps changes in demand, interest rates or inflation. Compounding by multiplying by one plus a proportion several times (raising to a power) is the key. For example, compound growth of 6% per annum over two years increases a sum of money by a factor of 1.06 × 1.06 = 1.1236. So $100 growing at 6% per annum for two years increases to $100 × 1.1236 = $112.36.

Proportions are also used in probability, which is worth looking at for its help in assessing risks.

Lastly, index numbers are introduced in this chapter.

Ways of looking at data
It is useful to be aware of different ways of categorising information. This is relevant for three main reasons.

1 *Time series and cross-sectional data*
Certain problems are found with some types of data only. For example, it is not too hard to see that you would look for seasonal and cyclical trends in time series but not in cross-sectional data.

Time series record developments over time; for example, monthly ice cream output, or a ten-year run of the finance director's annual salary.

Cross-sectional data are snapshots which capture a situation at a moment in time, such as the value of sales at various branches on one day.

2 Scales of measurement

Some techniques are used with one type of data only. A few of the sampling methods in Chapter 7 are used only with data which are measured on an interval or ratio scale. Other sampling methods apply to nominal or ordinal scale information only.

Nominal or categorical data identify classifications only. No particular quantities are implied. Examples include sex (male/female), departments (international/marketing/personnel) and sales regions (area number 1, 2, 3, 4).

Ordinal or ranked data. Categories can be sorted into a meaningful order, but differences between ranks are not necessarily equal. What do you think of this politician (awful, satisfactory, wonderful)? What grade of wheat is this (A1, A2, B1...)?

Interval scale data. Measurable differences are identified, but the zero point is arbitrary. Is 20° Celsius twice as hot as 10°C? Convert to Fahrenheit to see that it is not. The equivalents are 68°F and 50°F. Temperature is measured on an interval scale with arbitrary zero points (0°C and 32°F).

Ratio scale data. There is a true zero and measurements can be compared as ratios. If three frogs weigh 250gm, 500gm and 1,000gm, it is clear that Mr Frog is twice as heavy as Mrs Frog, and four times the weight of the baby.

3 Continuity

Some results are presented in one type of data only. You would not want to use a technique which tells you to send 0.4 of a salesman on an assignment, when there is an alternative technique which deals in whole numbers.

Discrete values are counted in whole numbers (integers): the number of frogs in a pond, the number of packets of Fat Cat Treats sold each week.

Continuous variables do not increase in steps. Measurements such as heights and weights are continuous. They can only be estimated: the temperature is 25°C; this frog weighs 500gm. The accuracy of such estimates depends on the precision of the measuring instrument. More accurate scales might show the weight of the frog at 501 or 500.5 or 500.0005 gm, etc.

Fractions, percentages and proportions

Fractions

Fractions are not complicated. Most monetary systems are based on 100 subdivisions: 100 cents to the dollar or euro, or 100 centimes to the Swiss franc. Amounts less than one big unit are fractions. 50 cents is half, or 0.50, or 50% of one euro. Common (vulgar) fractions (½), decimal fractions (0.50), proportions (0.50) and percentages (50%) are all the same thing with different names. Convert any common fraction to a decimal fraction by dividing the lower number (denominator) into the upper number (numerator). For example, ¾ = 3 ÷ 4 = 0.75. The result is also known as a proportion. Multiply it by 100 to convert it into a percentage. Recognition of these simple relationships is vital for easy handling of numerical problems.

Decimal places. The digits to the right of a decimal point are known as decimal places. 1.11 has two decimal places, 1.111 has three, 1.1111 has four, and so on.

Reading decimal fractions. Reading $10.45m as ten-point-forty-five million dollars will upset the company statistician. Decimal fractions are read out figure-by-figure: ten-point-four-five in this example. Forty-five implies four tens and five units, which is how it is to the left of

Percentage points and basis points

Percentages and percentage changes are sometimes confused. If an interest rate or inflation rate increases from 10% to 12%, it has risen by two units, or two percentage points. But the percentage increase is 20% (= 2 ÷ 10 × 100). Take care to distinguish between the two.

Basis points. Financiers attempt to profit from very small changes in interest or exchange rates. For this reason, one unit, say 1% (ie, one percentage point) is often divided into 100 basis points:

$$
\begin{aligned}
1 \text{ basis point} &= 0.01 \text{ percentage point} \\
10 \text{ basis points} &= 0.10 \text{ percentage point} \\
25 \text{ basis points} &= 0.25 \text{ percentage point} \\
100 \text{ basis points} &= 1.00 \text{ percentage point}
\end{aligned}
$$

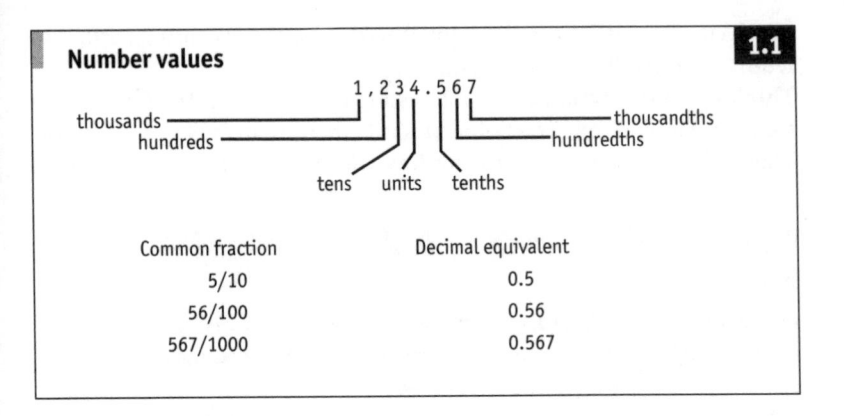

Number values **1.1**

Common fraction	Decimal equivalent
5/10	0.5
56/100	0.56
567/1000	0.567

Percentage increases and decreases

A percentage increase followed by the same percentage decrease does not leave you back where you started. It leaves you worse off. Do not accept a 50% increase in salary for six months, to be followed by a 50% cut.

◩ $1,000 increased by 50% is $1,500.
◩ 50% of $1,500 is $750.

A frequent business problem is finding what a number was before it was increased by a given percentage. Simply divide by $(1 + i)$, where i is the percentage increase expressed as a proportion. For example:

◩ if an invoice is for €575 including 15% VAT (value added tax, a sales tax) the tax-exclusive amount is €575 ÷ 1.15 = €500.

Fractions. If anything is increased by an amount x/y, the increment is $x/(x + y)$ of the new total:

◩ if €100 is increased by ½, the increment of €50 is $1/(1 + 2) = ⅓$ of the new total of €150;
◩ ¥100 increased by ¾ is ¥175; the ¥75 increment is $3/(3 + 4) = 3/7$ of the new ¥175 total.

the decimal point. To the right, the fractional amounts shrink further to tenths, hundredths, and so on. (See Figure 1.1.)

Think of two fractions. It is interesting to reflect that fractions go on for ever. Think of one fractional amount; there is always a smaller one. Think of two fractions; no matter how close together they are, there is

How big is a billion?

As individuals we tend to deal with relatively small amounts of cash. As corporate people, we think in units of perhaps one million at a time. In government, money seemingly comes in units of one billion only.

Scale. The final column below, showing that a billion seconds is about 32 years, gives some idea of scale. The fact that Neanderthal man faded away a mere one trillion seconds ago is worth a thought.

Quantity	Zeros	Scientific	In numbers	In seconds
Thousand	3	1×10^3	1,000	17 minutes
Million	6	1×10^6	1,000,000	11 ½ days
Billion	9	1×10^9	1,000,000,000	32 years
Trillion	12	1×10^{12}	1,000,000,000,000	32 thousand years

British billions. The number of zeros shown are those in common use. The British billion (with 12 rather than 9 zeros) is falling out of use. It is not used in this book.

Scientific notation. Scientific notation can be used to save time writing out large and small numbers. Just shift the decimal point along by the number of places indicated by the exponent (the little number in the air). For example:

- 1.25×10^6 is a shorthand way of writing 1,250,000;
- 1.25×10^{-6} is the same as 0.00000125.

Some calculators display very large or very small answers this way. Test by keying 1 ÷ 501. The calculator's display might show 1.996 −03, which means 1.996×10^{-3} or 0.001996. You can sometimes convert such displays into meaningful numbers by adding 1. Many calculators showing 1.996 −03 would respond to you keying + 1 by showing 1.00199. This helps identify the correct location for the decimal point.

always another one to go in between. This brings us to the need for rounding.

Rounding

An amount such as $99.99 is quoted to two decimal places when selling, but usually rounded to $100 in the buyer's mind. The Japanese have stopped counting their sen. Otherwise they would need wider calculators. A few countries are perverse enough to have currencies with three places of decimal: 1,000 fils = 1 dinar. But 1 fil coins are generally no longer in use and values such as 1.503 are rounded off to 1.505. How do you round 1.225 if there are no 5 fil coins? It depends whether you are buying or selling.

Generally, aim for consistency when rounding. Most calculators and spreadsheets achieve this by adopting the 4/5 principle. Values ending in 4 or less are rounded down (1.24 becomes 1.2), amounts ending in 5 or more are rounded up (1.25 becomes 1.3). Occasionally this causes problems.

Two times two equals four. Wrong: the answer could be anywhere between two and six when dealing with rounded numbers.

- 1.5 and 2.4 both round to 2 (using the 4/5 rule)
- 1.5 multiplied by 1.5 is 2.25, which rounds to 2
- 2.4 multiplied by 2.4 is 5.76, which rounds to 6

Also note that 1.45 rounds to 1.5, which rounds a second time to 2, despite the original being nearer to 1.

The moral is that you should take care with rounding. Do it after multiplying or dividing. When starting with rounded numbers, never quote the answer to more significant figures (see below) than the least precise original value.

Significant figures

Significant figures convey precision. Take the report that certain US manufacturers produced 6,193,164 refrigerators in a particular year. For some purposes, it may be important to know that exact number. Often, though, 6.2m, or even 6m, conveys the message with enough precision and a good deal more clarity. The first value in this paragraph is quoted to seven significant figures. The same amount to two significant figures is 6.2m (or 6,200,000). Indeed, had the first amount been estimated from refrigerator-makers' turnover and the average sale price of a refrigerator, seven-figure approximation would be spurious accuracy, of which economists are frequently guilty.

Significant figures and decimal places in use. Three or four significant figures with up to two decimal places are usually adequate for discussion purposes, particularly with woolly economic data. (This is sometimes called three or four effective figures.) Avoid decimals where possible, but do not neglect precision when it is required. Bankers would cease to make a profit if they did not use all the decimal places on their calculators when converting exchange rates.

Percentages and proportions

Percentages and proportions are familiar through money. 45 cents is 45% of 100 cents, or, proportionately, 0.45 of one dollar. Proportions are expressed relative to one, percentages in relation to 100. Put another way, a percentage is a proportion multiplied by 100. This is a handy thing to know when using a calculator.

Suppose a widget which cost $200 last year now retails for $220. Proportionately, the current cost is 1.1 times the old price (220 ÷ 200 = 1.1). As a percentage, it is 110% of the original (1.1 × 100 = 110).

In common jargon, the new price is 10% higher. The percentage increase (the 10% figure) can be found in any one of several ways. The most painless is usually to calculate the proportion (220 ÷ 200 = 1.1); subtract 1 from the answer (1.10 − 1 = 0.10); and multiply by 100 (0.10 × 100 = 10). Try using a calculator for the division and doing the rest by eye; it's fast.

Proportions and growth. The relationship between proportions and percentages is astoundingly useful for compounding.

The finance director has received annual 10% pay rises for the last ten years. By how much has her salary increased? Not 100%, but nearly 160%. Think of the proportionate increase. Each year, she earned 1.1 times the amount in the year before. In year one she received the base amount (1.0) times 1.1 = 1.1. In year two, total growth was 1.1 × 1.1 = 1.21. In year three, 1.21 × 1.1 = 1.331, and so on up to 2.358 × 1.1 = 2.594 in the tenth year. Take away 1 and multiply by 100 to reveal the 159.4 percentage increase over the whole period.

Powers. The short cut when the growth rate is always the same, is to recognise that the calculation involves multiplying the proportion by itself a number of times. In the previous example, 1.1 was multiplied by itself 10 times. In math-speak, this is called raising 1.1 to the power of 10 and is written 1.1^{10}.

The same trick can be used to "annualise" monthly or quarterly rates of growth. For example, a monthly rise in prices of 2.0% is equivalent to an annual rate of inflation of 26.8%, not 24%. The statistical section in the back of *The Economist* each week shows for 15 developed countries and the euro area the annualised percentage changes in output in the latest quarter compared with the previous quarter. If America's GDP is 1.7% higher during the January–March quarter than during the October–December quarter then this is equivalent to an annual rate of increase of 7% ($1.017 \times 1.017 \times 1.017 \times 1.017$).

Using a calculator. Good calculators have a key marked something like x^y, which means x (any number) raised to the power of y (any other number). Key 1.1 x^y 10 = and the answer 2.5937... pops up on the display. It is that easy. To go back in the other direction, use the $x^{1/y}$ key. So 2.5937 $x^{1/y}$ 10 = gives the answer 1.1. This tells you that the number that has to be multiplied by itself 10 times to give 2.5937 is 1.1. (See also Growth rates and exponents box, page 38.)

Table 1.1 **Mr & Mrs Average's shopping basket**

	A	B	C
	€	Jan 2000 = 100	Jan 2002 = 100
January 2000	1,568.34	100.0	86.6
January 2001	1,646.76	105.0	90.9
January 2002	1,811.43	115.5	100.0

Each number in column B = number in column A divided by (1,568.34 ÷ 100).
Each number in column C = number in column B divided by (1,811.43 ÷ 100).

Table 1.2 **A base-weighted index of living costs**

	A	B	C
	Food	Other	Total
Weights:	0.20	0.80	1.00
January 2000	100.0	100.0	100.0
January 2001	103.0	105.5	105.0
January 2002	108.0	117.4	115.5

Each monthly value in column C = (column A × 0.20) + (column B × 0.80).
Eg, for January 2002 (108.0 × 0.20) + (117.4 × 0.80) = 115.5.

Table 1.3 **A current-weighted index of living costs**

	A	B	C	D	E
	Food index	Food weight	Other index	Other weight	Total
January 2000	100.0	0.80	100.0	0.20	100.0
January 2001	103.0	0.70	105.5	0.30	103.8
January 2002	108.0	0.60	117.4	0.40	111.8

Each value in column E is equal to (number in column A × weight in column B) + (number in column C × weight in column D).
Eg, for January 2002 (108.0 × 0.60) + (117.4 × 0.40) = 111.8

Index numbers

There comes a time when money is not enough, or too much, depending on how you look at it. For example, the consumer prices index (also known as the cost of living or retail prices index) attempts to measure inflation as experienced by Mr and Mrs Average. The concept is straightforward: value all the items in the Average household's monthly shopping basket; do the same at some later date; and see how the overall cost has changed. However, the monetary totals, say €1,568.34 and €1,646.76 are not easy to handle and they distract from the task in hand. A solution is to convert them into index numbers. Call the base value 100. Then calculate subsequent values based on the percentage change from the initial amount. The second shopping basket cost 5% more, so the second index value is 105. A further 10% rise would take the index to 115.5.

To convert any series of numbers to an index:

- choose a base value (eg, €1,568.34 in the example here);
- divide it by 100, which will preserve the correct number of decimal places; then
- divide every reading by this amount.

Table 1.1 shows how this is done in practice.

Rebasing. To rebase an index so that some new period equals 100, simply divide every number by the value of the new base (Table 1.1).

Composite indices and weighting. Two or more sub-indices are often combined to form one composite index. Instead of one cost of living

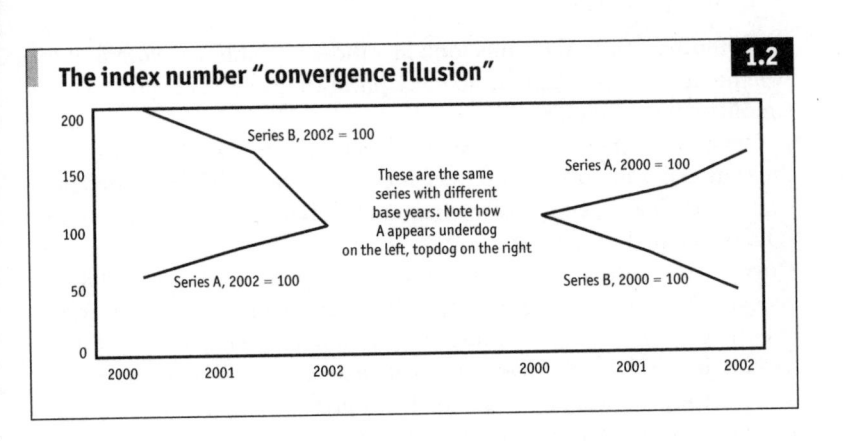

The index number "convergence illusion" **1.2**

Series B, 2002 = 100

These are the same series with different base years. Note how A appears underdog on the left, topdog on the right

Series A, 2000 = 100

Series A, 2002 = 100

Series B, 2000 = 100

Table 1.4 **Index comparisons**

	GDP per head $	Index USA = 100	Index UK = 100	Index Germany = 100
Japan	38,160	109.22	161.15	167.37
Norway	36,020	103.09	152.11	157.98
USA	34,940	100.00	147.55	153.25
Switrzerland	33,390	95.56	141.01	146.45
Denmark	30,420	87.06	128.46	133.42
Sweden	25,630	73.35	108.23	112.41
Ireland	24,740	70.81	104.48	108.51
UK	23,680	67.77	100.00	103.86
Finland	23,460	67.14	99.07	102.89
Austria	23,310	66.71	98.44	102.24
Netherlands	22,910	65.57	96.75	100.48
Germany	22,800	65.25	96.28	100.00
Canada	22,370	64.02	94.47	98.11
Belgium	22,110	63.28	93.37	96.97
France	21,980	62.91	92.82	96.40
Australia	20,340	58.21	85.90	89.21
Italy	18,620	53.29	78.63	81.67
Spain	14,150	40.50	59.76	62.06
New Zealand	13,030	37.29	55.03	57.15
Greece	10,670	30.54	45.06	46.80
Portugal	10,500	30.05	44.34	46.05

index for the Averages, there might be two: showing expenditure on food, and all other spending. How should they be combined?

Base weighting. The most straightforward way of combining indices is to calculate a weighted average. If 20% of the budget goes on food and

80% on other items, the sums look like those in Table 1.2. Note that the weights sum to one (they are actually proportions, not percentages); this simplifies the arithmetic.

Since this combined index was calculated using weights assigned at the start, it is known as a base-weighted index. Statisticians in the know sometimes like to call it a Laspeyres index, after the German economist who developed the first one.

Current weighting. The problem with weighted averages is that the weights sometimes need revision. With the consumer prices index, spending habits change because of variations in relative cost, quality, availability and so on. Indeed, UK statisticians came under fire as early as 1947 for producing an index of retail prices using outdated weights from a 1938 survey of family expenditure habits.

One way to proceed is to calculate a new set of current weights at regular intervals, and use them to draw up a long-term index. Table 1.3 shows one way of doing this.

This current-weighted index is occasionally called a Paasche index, again after its founder.

Imperfections and variations on weighting. Neither a base-weighted nor a current-weighted index is perfect. The base-weighted one is simple to calculate, but it exaggerates changes over time. Conversely, a current-weighted index is more complex to produce and it understates long-term changes. Neither Laspeyres nor Paasche got it quite right, and others have also tried and failed with ever more complicated formulae. Other methods to be aware of are Edgeworth's (an average of base and current weights), and Fisher's (a geometric average combining Laspeyres and Paasche indices).

Mathematically, there is no ideal method for weighting indices. Indeed, indices are often constructed using weights relating to some period other than the base or current period, or perhaps an average of several periods. Usually a new set of weights is introduced at regular intervals, maybe every five years or so.

Convergence. Watch for illusory convergence on the base. Two or more series will always meet at the base period because that is where they both equal 100 (see Figure 1.2). This can be highly misleading. Whenever you come across indices on a graph, the first thing you should do is check where the base is located.

Cross-sectional data. Index numbers are used not only for time series but also for snapshots. For example, when comparing salaries or other indicators in several countries, commentators often base their figures on

Summation and factorials

Summation. The Greek uppercase S, sigma or Σ, is used to mean nothing more scary than take the sum of. So "Σ profits" indicates "add up all the separate profits figures". Sigma is sometimes scattered with other little symbols to show how many of what to take the sum of. (It is used here only when this information is self-evident.) For example, to find the average salary of these four staff take the sum of their salaries and divide by four could be written: Σ salaries $\div$ 4.

Factorials. A fun operator is the factorial identified by an exclamation mark ! where 5! (read five factorial) means $5 \times 4 \times 3 \times 2 \times 1$. This is useful shorthand for counting problems.

their home country value as 100. This makes it easy to rank and compare the data, as Table 1.4 shows.

Notation

When you jot your shopping list or strategic plan on the back of an envelope you use shorthand. For example, "acq WW" might mean "acquire (or take over) World of Widgets". Sometimes you borrow from the world of numbers. Symbols such as + for add and = for equals need no explanation in the English-speaking world.

So why do we all freeze solid when we see mathematical shorthand? Not because it is hard or conceptually difficult, but because it is unfamiliar. This is odd. Mathematicians have been developing their science for a few thousand years. They have had plenty of time to develop and promulgate their abbreviations. Some are remarkably useful, making it simple to define problems concisely, after which the answer is often self-evident. For this reason, this book does not fight shy of symbols, but they are used only where they aid clarity.

The basic operators $+ - \times \div$ are very familar. Spreadsheet and other computer users will note that to remedy a keyboard famine $\div$ is replaced by / and $\times$ is replaced by $\star$. Mathematicians also sometimes use / or write one number over another to indicate division, and omit the multiplication sign. Thus if $a = 6$ and $b = 3$:

$$a \div b = a/b = \tfrac{a}{b} = \tfrac{a}{b} = 2 \text{ and}$$
$$a \times b = ab = a \cdot b = 18$$

Brackets. When the order of operation is important, it is highlighted with brackets. Perform operations in brackets first. For example, $4 \times (2 + 3) = 20$ is different from $(4 \times 2) + 3 = 11$. Sometimes more than one set of brackets is necessary, such as in $[(4 \times 2) + 3] \times 6 = 66$. When entering complex formulas in spreadsheet cells, always use brackets to ensure that the calculations are performed as intended.

Powers. When dealing with growth rates (compound interest, inflation, profits), it is frequently necessary to multiply a number by itself many times. Writing this out in full becomes laborious. To indicate, for example, $2 \times 2 \times 2 \times 2$, write 2^4 which is read "two raised to the power of four". The little number in the air is called an exponent (Latin for out-placed).

Roots. Just as the opposite of multiplication is division, so the opposite of raising to powers is taking roots. $\sqrt[4]{625}$ is an easy way to write "take the fourth root of 625" – or in this case, what number multiplied by itself four times equals 625? (Answer 5, since $= 5 \times 5 \times 5 \times 5 = 625$). The second root, the square root, is generally written without the 2 (eg, $\sqrt[2]{9} = \sqrt{9} = 3$). Just to confuse matters, a convenient alternative way of writing "take a root" is to use one over the exponent. For example, $16^{1/4} = \sqrt[4]{16} = 2$.

Equalities and inequalities. The equals sign = (or equality) needs no explanation. Its friends, the inequalities, are also useful for business problems. Instead of writing "profits must be equal to or greater than ¥5m", scribble "profits $\geq$ ¥5m". Other inequalities are:

- less than or equal to $\leq$;
- greater than >; and
- less than <.

They are easy to remember since they open up towards greatness. A peasant < a prince (perhaps). Along the same lines not equal $\neq$ and approximately equal $\approx$ are handy.

Symbols

Letters such as a, b, x, y and n sometimes take the place of constants or variables – things which can take constant or various values for the purpose of a piece of analysis.

For example, a company trading in xylene, yercum and zibeline (call these x, y and z), which it sells for €2, €3 and €4 a unit, would calculate sales revenue (call this w) as:

$$w = (2 \times x) + (3 \times y) + (4 \times z)$$
$$\text{or } w = 2x + 3y + 4z$$

When sales figures are known at the end of the month, the number of units sold of xylene, yercum and zibeline can be put in place of x, y and z in the equation so that sales revenue w can be found by simple arithmetic. If sales prices are as yet undetermined, the amounts shown above as €2, €3 and €4 could be replaced by a, b and c so that the relationship between sales and revenue could still be written down:

$$w = (a \times x) + (b \times y) + (c \times z)$$
$$\text{or } w = ax + by + cz$$

See below for ways of solving such equations when only some of the letters can be replaced by numerical values.

When there is a large number of variables or constants, there is always a danger of running out of stand-in letters. Alternative ways of rewriting the above equation are:

$$w = (a \times x_1) + (b \times x_2) + (c \times x_3)$$
$$\text{or even } x_0 = (a_1 \times x_1) + (a_2 \times x_2) + (a_3 \times x_3)$$

The little numbers below the line are called subscripts, where x_1 = xylene, x_2 = yercum, and so on.

General practice. There are no hard and fast rules. Lowercase letters near the beginning of the alphabet (a, b, c) are generally used for constants, those near the end (x, y, z) for variables. Frequently, y is reserved for the major unknown which appears on its own on the left-hand side of an equation, as in $y = a + (b \times x)$. The letter n is often reserved for the total number of observations, as in "we will examine profits over the last n months" where n might be set at 6, 12 or 24.

Romans and Greeks. When the Roman alphabet becomes limiting, Greek letters are called into play. For example, in statistics, Roman letters are used for sample data (p = proportion from a sample). Greek equivalents indicate population data (π indicates a proportion from a population: see page 126).

Circles and pi. To add to the potential confusion the Greek lower-case p (π, pi) is also used as a constant. By a quirk of nature, the distance around the edge of a circle with a diameter of 1 foot is 3.14 feet. This measurement is important enough to have a name of its own. It is labelled π, or pi. That is, $\pi = 3.14$. Interestingly, pi cannot be calculated exactly. It is 3.1415927 to eight significant figures. It goes on forever and is known as an irrational number.

Solving equations

Any relationship involving an equals sign = is an equation. Two examples of equations are $3 + 9x = 14$ and $(3 \times x) + (4 \times y) = z$. The following three steps will solve any everyday equation. They may be used in any order and as often as necessary.

1 **Add or multiply.** The equals sign = is a balancing point. If you do something to one side of the equation you must do the same to the other.

Addition and subtraction

With this equation subtract 14 from both sides to isolate y:

$$y + 14 = x$$
$$(y + 14) - 14 = x - 14$$
$$y = x - 14$$

Multiplication and division

With this equation divide both sides by 2 to isolate y:

$$y \times 2 = x$$
$$(y \times 2) \div 2 = x \div 2$$
$$y = x \div 2$$

2 **Remove awkward brackets.**

$$2 \times (6 + 8) = (2 \times 6) + (2 \times 8) = 12 + 16 = 28$$
$$2 \times (x + y) = (2 \times x) + (2 \times y)$$

3 **Dispose of awkward subtraction or division.**

Subtraction is negative addition (eg, $6 - 4 = 6 + -4 = 2$).

Note that a plus and a minus is a minus (eg, $6 + -2 = 4$), while two minus signs make a plus (eg, $6 - -2 = 8$).

Division by x is multiplication by the reciprocal $\frac{1}{x}$ (eg, $6 \div 3 = 6 \times \frac{1}{3} = 2$).

Note that $3 \times \frac{1}{3} = 3 \div 3 = 1$.

For example, suppose that in the following relationship, y is to be isolated:

$$^w/_6 = (2 \times y) + (12 \times x) - 3$$

Subtract $(12 \times x)$ from both sides:

$$^w/_6 - (12 \times x) = (2 \times y) + (12 \times x) - (12 \times x) - 3$$
$$^w/_6 - (12 \times x) = (2 \times y) - 3$$

Add 3 to both sides:

$$^w/_6 - (12 \times x) + 3 = (2 \times y) - 3 + 3$$
$$^w/_6 - (12 \times x) + 3 = (2 \times y)$$

Multiply both sides by ½:

$$½ \times [^w/_6 - (12 \times x) + 3] = ½ \times (2 \times y)$$
or $^w/_{12} - (6 \times x) + ^3/_2 = y$
ie, $y = ^w/_{12} - (6 \times x) + ^3/_2$

Probability
Rationalising uncertainty
Uncertainty dominates business (and other) decisions. Yet there is a certainty about uncertainty that makes it predictable; it is possible to harness probability to improve decision-making.

For example, right now, someone might be stealing your car or burgling your home or office. If you are not there, you cannot be certain, nor can your insurance company. It does not know what is happening to your neighbour's property either, or to anyone else's in particular. What the insurance company does know from experience is that a given number (1, 10, 100 ...) of its clients will suffer a loss today.

Take another example: toss a coin. Will it land heads or tails? Experience or intuition suggests that there is a 50:50 chance of either. You cannot predict the outcome with certainty. You will be 100% right or 100% wrong. But the more times you toss it, the better your chance of predicting in advance the proportion of heads. Your percentage error tends to reduce as the number of tosses increases. If you guess at 500 heads before a marathon 1,000-toss session, you might be only a fraction of a percent out.

The law of large numbers. In the long run, the proportion (relative frequency) of heads will be 0.5. In math-speak, probability is defined as the limit of relative frequency. The limit is reached as the number of repetitions approaches infinity, by which time the proportion of heads should be fairly and squarely at 0.500. The tricky part is that you can never quite get to infinity. Think of a number – you can always add one more. It is similar to the "think of two fractions" problem mentioned earlier. Mathematicians are forced to say it is probable that the limit of relative frequency will be reached at infinity. This is called the law of large numbers, one of the few theorems with a sensible name. It seems to involve circular reasoning (probability probably works at infinity). But there is a more rigorous approach through a set of laws (axioms) which keeps academics happy.

Applications. In many cases, probability is helpful for itself. Quantifying the likelihood of some event is useful in the decision-making process. But probability also forms the basis for many very interesting decision-making techniques discussed in the following chapters. The ground rules are considered below.

Estimating probabilities

Measuring. Everyday gambling language (10 to 1 odds on a horse, a 40% chance of rain) is standardised in probability-speak.

Probability is expressed on a sliding scale from 0 to 1. If there is no chance of an event happening, it has a probability of zero. If it must occur, it has a probability of 1 (this is important). An unbiased coin can land on a flat surface in one of only two ways. There is a 50% chance of either. Thus, there is a 0.5 probability of a head, and a 0.5 probability of a tail. If four workers are each equally likely to be selected for promotion, there is a 1 in 4, or 25%, chance that any one will be selected. They each have a 0.25 probability of rising further towards their level of incompetence. Probabilities are logic expressed proportionately.

Certainty = 1. Why highlight the importance of an unavoidable event having a probability of one? Look at the coin example. There is a 0.5 probability of a head and a 0.5 probability of a tail. One of these two events must happen (excluding the chances of the coin staying in the air or coming to rest on its edge), so the probabilities must add up to one.

If the probability of something happening is known, then by definition the probability of it not happening is also known. If the meteorologist's new computer says there is a 0.6 probability of rain tomorrow, then there is a 0.4 probability that it will not rain. (Of course, the meteorologist's computer might be wrong.)

Assigning probabilities using logic. When the range of possible outcomes can be foreseen, assigning a probability to an event is a matter of

Probability rules

The probability of any event P(A) is a number between 0 and 1, where 1 = certainty.

GENERAL RULE

The probability of event A is the number of outcomes where A happens n_A divided by the total number of possible outcomes n

$P(A) = n_A \div n.$

The probability of an event not occurring is equal to the one minus the probability of it happening

$P(\text{not } A) = 1 - P(A)$

PROBABILITY OF COMPOSITE EVENTS

One outcome given another

The probability of event A given that event B is known to have occurred = number of outcomes where A happens when A is selected from B (n_A) divided by number of possible outcomes B (n_B)

$P(A|B) = n_A \div n_B$

One outcome or the other [b]

Independent (mutually exclusive) events

$P(A \text{ or } B) = P(A) + P(B)$

Dependent (non-exclusive) events

$P(A \text{ or } B) = P(A) + P(B) - P(A \text{ and } B)$

Both outcomes [b]

Independent (mutually exclusive) events

$P(A \text{ and } B) = P(A) \times P(B)$

Dependent (non-exclusive) events

$P(A \text{ and } B) = P(A) \times P(B|A)$

or

$P(A \text{ and } B) = P(B) \times P(A|B)$

EXAMPLE

P(red ace), $P(A) = 2/52 = 0.038$[a]

P(any heart), $P(B) = 13/52 = 0.25$

P(king of clubs), $P(C) = 1/52 = 0.019$

Probability of drawing at random any card other than a red ace =

$P(\text{not } A) = 1 - 2/52 = 50/52 = 0.962$

Probability that the card is a red ace given that you know a heart was drawn =

$P(A|B) = 1/13 = 0.077$

Probability of drawing a red ace or any club =

$P(A \text{ or } B) = 2/52 + 13/52 = 15/52$

Probability of drawing the king of clubs or any club =

$P(A \text{ or } B) = 1/52 + 13/52 - 1/52 = 13/52$

Probability of drawing a red ace, returning it to the pack and then drawing any heart = $P(A \text{ and } B) = 2/52 \times 13/52 = 1/104$

Probability of drawing one card which is both a red ace and a heart = $P(A \text{ and } B) = 2/52 \times 1/2 = 1/52$ or $P(A \text{ and } B) = 13/52 \times 1/13 = 1/52$

a There are 52 cards in a pack, split into four suits of 13 cards each. Hearts and diamonds are red, while the other two suits, clubs and spades, are black.

b If in doubt, treat events as dependent.

simple arithmetic. Reach for the coin again. Say you are going to toss it three times. What is the probability of only two heads? The set of all possible outcomes is as follows (where, for example, one outcome from three tosses of the coin is three heads, or HHH):

HHH   THH   HTH   HHT   TTH   HTT   THT   TTT

Of the eight equally possible outcomes, only three involve two heads. There is a 3 in 8 chance of two heads. The probability is ⅜, or 0.375. Look at this another way. Each outcome has a ⅛ = 0.125 chance of happening, so the probability of two heads can also be found by addition: 0.125 + 0.125 + 0.125 = 0.375.

The likelihood of not getting two heads can be computed in one of three ways. Either of the two approaches outlined may be used. But a simple method is to remember that since probabilities must sum to one, failure to achieve two heads must be 1 − 0.375 = 0.625.

This has highlighted two important rules.

- If there are a outcomes where event A occurs and n outcomes in total, the probability of event A is calculated as a ÷ n. The shorthand way of writing this relationship is P(A) = a ÷ n.
- The probability of an event not occurring is equal to the probability of it happening subtracted from one. In shorthand: P(not A) = 1 − P(A).

Assigning probabilities by observation. When probabilities cannot be estimated using the foresight inherent in the coin-tossing approach, experience and experiment help. If you know already that in every batch of 100 widgets 4 will be faulty, the probability of selecting a wobbly widget at random is 4 ÷ 100 = 0.04. If 12 out of every 75 shoppers select Fat Cat Treats, there is a 12 ÷ 75 = 0.16 probability that a randomly selected consumer buys them.

Subjective probabilities. On many occasions, especially with business problems, probabilities cannot be found from pure logic or observation. In these circumstances, they have to be allocated subjectively. You might, for example, say "considering the evidence, I think there is a 10% chance (ie, a 0.10 probability) that our competitors will imitate our new product within one year". Such judgments are acceptable as the best you can do when hard facts are not available.

1.3

Multiple events

The probability of a complex sequence of events can usually be found by drawing a little tree diagram. For example, the following tree shows the probability of selecting hearts on two consecutive draws (ie, without replacement) from a pack of 52 cards. Starting at the left, there are two outcomes to the first draw, a heart or a non-heart. From either of these two outcomes, there are two further outcomes. The probability of each outcome is noted, and the final probabilities are found by the multiplication rule. Check the accuracy of your arithmetic by noting that the probabilities in the final column must add up to one, since one of these outcomes must happen.

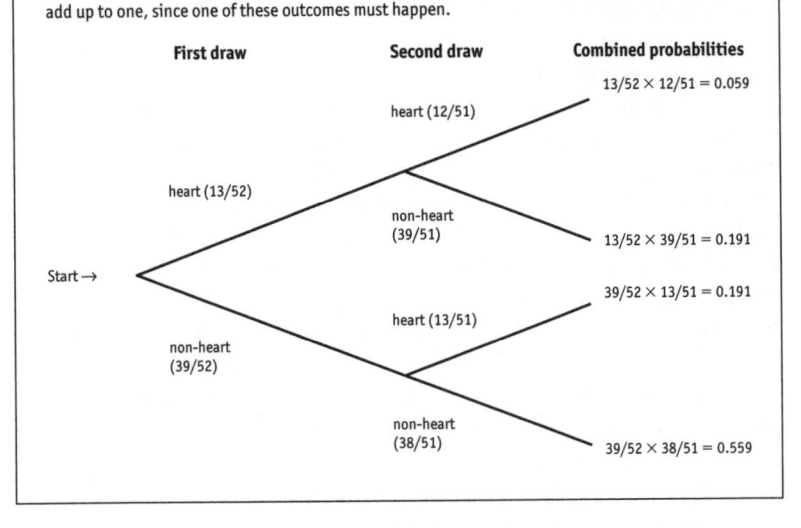

First draw	Second draw	Combined probabilities

heart (12/51) $13/52 \times 12/51 = 0.059$

heart (13/52)

non-heart (39/51) $13/52 \times 39/51 = 0.191$

Start →

heart (13/51) $39/52 \times 13/51 = 0.191$

non-heart (39/52)

non-heart (38/51) $39/52 \times 38/51 = 0.559$

Composite events

There are some simple rules for deriving the probabilities associated with two or more events. You might know the risks of individual machines failing and want to know the chances of machine A or B failing; or the risk of both A and B breaking down at the same time.

The basic rules are summarised in the box on page 23. The following examples of how the rules work are based on drawing cards from a pack of 52 since this is relatively easy to visualise.

Composite events: A given B

Sometimes you want to know probabilities when you have some advance information. For example, if warning lights 4 and 6 are flashing, what are the chances that the cause is a particular malfunction? The solution is found by logically narrowing down the possibilities. This is easy to see with the playing card example.

The question might be phrased as "What is the probability that the selected card is a king given that a heart was drawn?". The shorthand notation for this is P(K|H), where the vertical bar is read "given". Since there are 13 hearts and only one king of hearts the answer must be P(K|H) = $\frac{1}{13}$ = 0.077. This is known as conditional probability, since the card being a king is conditional on having already drawn a heart.

Revising probabilities as more information becomes available is considered in Chapter 8.

Composite events: A or B

Add probabilities to find the chance of one event or another (eg, what are the chances of order A or order B arriving at a given time?).

Mutually exclusive events. The probability of drawing a heart or a black king is $\frac{13}{52}$ + $\frac{2}{52}$ = 0.288. Note that the two outcomes are mutually exclusive because you cannot draw a card which is both a heart and a black king. If a machine can process only one order at a time, the chances that it is dealing with order A or order B are mutually exclusive.

Non-exclusive events. The probability of drawing a heart or a red king is not $\frac{13}{52}$ + $\frac{2}{52}$ = 0.288. This would include double counting. Both the "set of hearts" and the "set of red kings" include the king of hearts. It is necessary to allow for the event which is double counted (the probability of drawing a heart and a red king). This is often done most easily by calculating the probability of the overlapping event and subtracting it from the combined probabilities. There is only one card which is both a king and a heart, so the probability of this overlapping event is $\frac{1}{52}$. Thus, the probability of drawing a heart or a red king is $\frac{13}{52}$ + $\frac{2}{52}$ − $\frac{1}{52}$ = 0.269. If a machine can process several orders at a time, the chances that it is dealing with order A or order B are not mutually exclusive.

Composite events: A and B

Multiply probabilities together to find the chance of two events occurring simultaneously (eg, receiving a large order and having a vital machine break down on the same day).

Independent events. The probability of drawing a red king from one pack and a heart from another pack is $\frac{2}{52}$ × $\frac{13}{52}$ = 0.01. These two events are independent so their individual probabilities can be multiplied together to find the combined probability. A machine breakdown and receipt of a large order on the same day would normally be independent events.

Dependent events. Drawing from a single pack, a card which is both a red king and a heart is a dependent composite event. The probability is

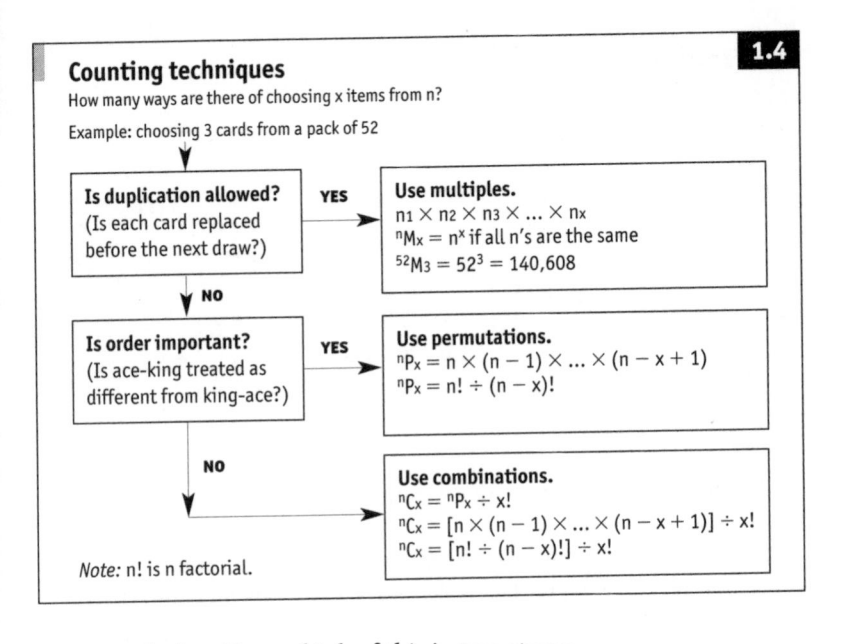

Counting techniques

How many ways are there of choosing x items from n?

Example: choosing 3 cards from a pack of 52

Is duplication allowed? (Is each card replaced before the next draw?) — **YES** → **Use multiples.**
$n_1 \times n_2 \times n_3 \times ... \times n_x$
$^nM_x = n^x$ if all n's are the same
$^{52}M_3 = 52^3 = 140,608$

NO

Is order important? (Is ace-king treated as different from king-ace?) — **YES** → **Use permutations.**
$^nP_x = n \times (n-1) \times ... \times (n-x+1)$
$^nP_x = n! \div (n-x)!$

NO

Use combinations.
$^nC_x = {^nP_x} \div x!$
$^nC_x = [n \times (n-1) \times ... \times (n-x+1)] \div x!$
$^nC_x = [n! \div (n-x)!] \div x!$

Note: n! is n factorial.

easy to calculate if you think of this in two stages.

- ☑ What is the probability of drawing a red king (²⁄₅₂)?
- ☑ Given that the card was a red king, what is the (conditional) probability that it is also a heart (½)?

These two probabilities multiplied together give the probability of a red king and a heart, ²⁄₅₂ × ½ = ¹⁄₅₂. This answer can be verified by considering that there is only one card which meets both conditions so the probability must be ¹⁄₅₂. A machine breakdown might be dependent on the receipt of a large order if the order overloads the machine. It is often difficult to decide whether events are independent or dependent. If in doubt, treat them as dependent.

Counting techniques

A frequent problem with probability is working out how many events are actually taking place. Visualise a card game. What are the chances that a five-card hand will contain four aces? It is the number of five-card hands containing four aces, divided by the total number of possible five-card hands. The two numbers that go into solving this equation are slightly elusive. So are the numbers required to solve some business

Ten factorials

0! =	1 =	1
1! =	1 =	1
2! =	$2 \times 1 =$	2
3! =	$3 \times 2 \times 1 =$	6
4! =	$4 \times 3 \times 2 \times 1 =$	24
5! =	$5 \times 4 \times 3 \times 2 \times 1 =$	120
6! =	$6 \times 5 \times 4 \times 3 \times 2 \times 1 =$	720
7! =	$7 \times 6 \times 5 \times 4 \times 3 \times 2 \times 1 =$	5,040
8! =	$8 \times 7 \times 6 \times 5 \times 4 \times 3 \times 2 \times 1 =$	40,320
9! =	$9 \times 8 \times 7 \times 6 \times 5 \times 4 \times 3 \times 2 \times 1 =$	362,880

problems, such as "how many different ways are there that I can fulfil this order?".

There are three counting techniques (multiples, permutations and combinations – see Figure 1.4) which solve nearly all such problems.

Multiples

The multiples principle applies where duplication is permitted. A health clinic gives its patients identification codes (IDs) comprising two letters followed by three digits. How many IDs are possible? There are 26 characters that could go into the first position, 26 for the second, 10 for the third, 10 for the fourth, and 10 for the fifth. So there must be $26 \times 26 \times 10 \times 10 \times 10 = 676{,}000$ possible IDs.

Think of each position as a decision. There are 26 ways to make the first decision, 26 ways to make the second decision, and so on through to ten ways to make the fifth decision. This may be generalised as n_1 ways to make decision D_1, n_2 ways to make D_2, n_3 ways to make D_3 and so on. Thus, there are $n_1 \times n_2 \times n_3 \times \ldots \times n_x$ ways to make x decisions.

Powers. There is a handy shortcut for situations where there are the same number of options for each decision. How many sequences are generated by nine tosses of a coin? (HTHHTTTHH is one sequence.) There are two ways for each decision: $2 \times 2 \times 2 \times 2 \times 2 \times 2 \times 2 \times 2 \times 2 = 2^9 = 512$. In this case, the multiples principle may be written $^nM_x = n^x$ for short, where ^nM_x is read as the number of multiples of x items selected from a total of n.

Combinations and permutations
Arrangements of the 1 (ace), 2, 3 and 4 of hearts

Combinations			Permutations			
1,2,3	1,2,3	1,3,2	2,1,3	2,3,1	3,1,2	3,2,1
1,2,4	1,2,4	1,4,2	2,1,4	2,4,1	4,1,2	4,2,1
1,3,4	1,3,4	1,4,3	3,1,4	3,4,1	4,1,3	4,3,1
2,3,4	2,3,4	2,4,3	3,2,4	3,4,2	4,2,3	4,3,2

←→   ←――――――――――――――――――――――――→

↑   ↑

3 items in each row $3! = 3 \times 2 \times 1 = 6$ permutations for each set of 3.

Permutations

With multiples the same values may be repeated in more than one position. Permutations are invoked when duplication is not allowed. For example, four executives are going to stand in a row to be photographed for their company's annual report. Any one of the four could go in the first position, but one of the remaining three has to stand in the second position, one of the remaining two in the third position, and the remaining body must go in the fourth position. There are $4 \times 3 \times 2 \times 1 = 24$ permutations or 24 different ways of arranging the executives for the snapshot.

Factorials. As previewed above, declining sequences of multiplication crop up often enough to be given a special name: factorials. This one ($4 \times 3 \times 2 \times 1$) is 4 factorial, written as 4! Good calculators have a key for calculating factorials in one move. The first ten are shown in the box on page 28. Incidentally, mathematical logic seems to be wobbly with 0! which is defined as equal to 1.

Permutations often involve just a little more than pure factorials. Consider how many ways there are of taking three cards from a pack of 52. There are 52 cards that could be selected on the first draw, 51 on the second and 50 on the third. This is a small chunk of 52 factorial. It is 52! cut off after three terms. Again, this can be generalised. The number of permutations of n items taken x at a time is $^nP_x = n \times (n - 1) \dots (n - x + 1)$. The final term is ($52 - 3 + 1 = 50$) in the three-card example. This reduces to $52! \div 49!$ or $^nP_x = n! \div (n - x)!$ in general; which is usually the most convenient way to deal with permutations:

$$^{52}P_3 = \frac{52 \times 51 \times 50 \times 49 \times 48 \times \dots \times 2 \times 1}{49 \times 48 \times \dots \times 2 \times 1}$$

Combinations

Combinations are used where duplication is not permitted and order is not important. This occurs, for example, with card games; dealing an ace and then a king is treated as the same as dealing a king followed by an ace. Combinations help in many business problems, such as where you want to know the number of ways that you can pick, say, ten items from a production batch of 100 (where selecting a green Fat Cat Treat followed by a red one is the same as picking a red followed by a green).

To see how the combinations principle works, consider four playing cards, say the ace (or 1), 2, 3 and 4 of hearts.

- ◪ The four combinations of three cards that can be drawn from this set are shown on the left-hand side of Figure 1.5.
- ◪ The six possible permutations of each combination are shown on the right-hand side of Figure 1.5.
- ◪ For each set of 3 cards, there are $3! = 3 \times 2 \times 1 = 6$ permutations, or in general for each set of x items there are $x!$ permutations.
- ◪ So the total number of permutations must be the number of combinations multiplied by $x!$, or $4 \times 3! = 24$. In general purpose shorthand, $^{n}P_{x} = {^{n}C_{x}} \times x!$
- ◪ This rearranges to $^{n}C_{x} = {^{n}P_{x}} \div x!$; the relationship used to find the number of combinations of x items selected from n. In other words, when selecting 3 cards from a set of 4, there are $24 \div 3! = 4$ possible combinations.

The card puzzle

The puzzle which introduced this section on counting techniques asked what the chances were of dealing a five-card hand which contains four aces. You now have the information required to find the answer.

1 How many combinations of five cards can be drawn from a pack of 52? In this case, $x = 5$ and $n = 52$. So $^{52}C_{5} = [52! \div (52 - 5)!] \div 5! = 2,598,960$. There are 2.6m different five-card hands that can be dealt from a pack of 52 cards.
2 How many five-card hands can contain four aces? These hands must contain four aces and one other card. There are 48 non-aces, so there must be 48 ways of dealing the four-ace hand:

$$^{4}C_{4} \times {^{48}C_{1}} = [(4! \div 0!) \div 4!] \times [(48! \div 47! \div 1!] = 48.$$

3 The probability of dealing a five-card hand containing exactly four aces is $^{48}/2{,}598{,}960 = 0.0000185$ or 1 in 54,145 – not something to put money on.

Encryption

There is one more topic which we might consider under Key Concepts: encryption. This relies on mathematical algorithms to scramble text so that it appears to be complete gibberish to anyone without the *key* (essentially, a password), to unlock or decrypt it. You do not need to understand encryption to use it, but it is an interesting topic which deserves at least a passing mention here. There are two key techniques:

- **Symmetric encryption** uses the same key to encrypt and decrypt messages.
- **Asymmetric encryption** involves the use of multiple keys (it is also known as dual-key or public-key encryption). Each party to an encrypted message has two keys, one held privately and the other known publicly. If I send you a message, it is encrypted using my private key and your public key. You decrypt it using your private key.

For a given amount of processing power, symmetric encryption is stronger than the asymmetric method, but if the keys have to be transmitted the risk of compromise is lower with asymmetric encryption. It is harder to guess or *hack* keys when they are longer. A 32-bit key (where each bit is 0 or 1) has 2^{32} or over 4 billion combinations. Depending on the encryption technique and the application, 128-bit and 1024-bit keys are generally used for Internet security.

To provide a flavour of the mathematics involved, consider the most famous asymmetric encryption algorithm, RSA (named for its inventors Rivest, Shamir, and Adleman). It works as follows:

1 Take two large prime numbers, p and q, and compute their product $n = pq$ (a prime number is a positive integer greater than 1 that can be divided evenly only by 1 and itself).
2 Choose a number, e, less than n and relatively prime to $(p-1)(q-1)$, which means e and $(p-1)(q-1)$ have no common factors except 1.
3 Find another number d such that $(ed-1)$ is divisible by $(p-1)(q-1)$.

The values e and d are called the public and private exponents, respectively. The public key is the pair (n, e); the private key is (n, d). Key length, as discussed above, relates to n, which is used as a modulus (a divisor where the result of integer division is discarded and the remainder is kept – see MOD in the A–Z).

The characters of the alphabet are assigned numerical values. I can send you a *ciphertext* c of any character value m, using your public key (n, e); where $c = m^e \bmod n$. To decrypt, you exponentiate $m = c^d \bmod n$. The relationship between e and d ensures that you correctly recover m. Since only you know d, only you can decrypt this message. Hackers note: it is difficult to obtain the private key d from the public key (n, e), but if you *could* factor n into p and q, then you could obtain the private key d.

2 Finance and investment

"A banker is a man who lends you an umbrella when the
weather is fair, and takes it away from you when it rains."

Anon

Summary

Problems which appear to relate specifically to money usually involve
growth over time. Growth results from interest accruals, inflation, or
(more usually) both.

The most common money problem is dealing with investments
which produce differing returns over different time periods. The word
"investment" is used loosely here to mean either fixed investment in
capital projects or savings investments. In terms of numerical analysis,
the two are much the same. Usually the objective is to find the highest
yielding investment. Since one person's outflow is another's inflow,
exactly the same arithmetic also relates to finding the cheapest loan.

In its simplest form, the problem reduces to one of measuring alternatives against a standard yardstick; for example, a completely safe
interest-bearing bank account. An investment is only worthwhile if it
promises a higher return than the bank.

The arithmetic of interest rates is the key to comparing investment
opportunities, financial or capital. The two main approaches, known
collectively as discounted cash flow (DCF) techniques, are net present
value (what sum of money would have to be banked today to produce
the same cash flow as this project?) and internal rate of return (if this
project were a bank deposit, what rate of interest would it be paying?).

The same arithmetic is used when pricing rental and lease contracts,
deciding whether to rent or buy, and deciding whether to settle immediately for discount or pay later.

The final money problem discussed in this chapter is that of
exchange rates where understanding the terminology is the most important aspect.

Interest

Interest is the price paid for the use of a sum of money. This is exactly the same as a charge for renting a car or accommodation. The longer the item is out of its owner's hands, the greater the fee. Most rental charges are quoted as so much cash per unit per charging period: $500 per car per week; €5,000 per truck per month; ¥5m per office per year. So it is with money. For example, $5 per $100 per charging period, or 5 per cent per period. The charge is generally standardised as a percentage (ie, %, or per hundred) interest rate. The period on which the rate is based tends to be standardised at one year (eg, 5% a year) regardless of the length of the actual rental (the term) – which is a completely separate matter. If the charging period is not quoted it is probably one year, although this is not necessarily so.

Borrowers and lenders
Every transaction has two sides. If you borrow money from Universal Credit, Universal is the lender and you are the borrower. You receive a loan which you repay in stages with interest charged on the outstanding balance. Universal sees it as a lump sum investment which earns interest and which is paid back in instalments. Similarly, a bank deposit should also be seen as a loan to the bank. If an interest rate problem seems difficult to solve, look at it from the other angle.

Simple interest
Simple interest is calculated on only the principal, the original capital sum. For example, with $100 invested in a 6% income fund, payments of $6 a year are remitted to the investor while the capital sum remains unchanged. At the end of five years the principal is still $100 and the total accumulated interest is $30. In interest rate-speak the money's present value is $100. Its future value, the amount it will grow to in five years with this investment, is $130.

The obvious trap with simple interest is that the interest payments may be reinvested elsewhere to increase the principal's overall future value. Unless reinvestment details are known, sensible comparison of simple interest rates may be impossible.

For example, which is better: $10,000 in a bank deposit for six months at 10% a year, or for one year at 6%? The six-month plan earns $500 in interest, the longer one generates $600. But there is not enough information to answer the question sensibly.

Possibly the shorter six-month investment is better for the following reasons. First, the investor's money is returned more rapidly. Most investors have a time preference for money. The sooner they have it in their hands, the sooner they can choose what to do with it next. Second,

Simple interest

General rule	*Example*
Where	
PV = present value (principal sum)	$1,000
i = annual rate of interest as proportion	0.125 (ie, 12.5%)
t = number of years	5
Interest factor r = i × t	= 0.125 × 5 = 0.625
Future value FV = PV × (1 + r)	= $1,000 × (1 + 0.625)
	= $1,000 × 1.625
	= $1,625
Simple interest SI = PV × r	= $1,000 × 0.625 = $625
or SI = FV − PV	= $1,625 − 1,000 = $625
Simple interest rate i = (SI ÷ PV) ÷ t	= ($625 ÷ 1,000) ÷ 5
	= 0.125 (ie, 12.5%)

the risks of never seeing the money again are reduced. The longer it is elsewhere (in a bank, the stock market, etc) the higher the chances of some financial crisis. Third, the cash could be reinvested for a second six-month period and earn more than an extra $100.

However, investors might choose the lower-yielding longer-term plan if they think that the bank is totally safe, are in no hurry to have back the $10,000, and are pessimistic about finding a suitable reinvestment after the first six months.

Simple interest arithmetic. A proportion is to 1 what a percentage is to 100 (see pages 12ff). For example, 6% is the same as 0.06. So 6% of $100 = $100 × 0.06 = $6.

Think of this 0.06 as the interest factor. For more than one time period, the interest factor equals the interest rate multiplied by time. For a five-year investment at a simple interest rate of 6% a year, the interest factor is 5 × 0.06 = 0.30. So $100 invested for five years at a simple interest rate of 6% a year, earns interest of $100 × 0.30 = $30.

Future value. Future value equals principal plus interest. This can be written as future value = (principal × 1) + (principal × interest factor) or more neatly as future value = principal × (1 + interest factor).

For example, $100 invested for five years at a simple interest rate of 6%, grows to a future value of $100 × 1.30 = $130.

Interest accumulation factor. The expression (1 + interest factor), the interest accumulation factor if you like, is the key to successful interest rate calculations. It is usually quoted in shorthand as (1 + r) where r is the interest factor (see Simple interest box, page 35).

Treat parts of a year as fractions. Simple interest of 6% on $100 borrowed for a year and a half generates a future value of:

$$\$100 \times (1 + [0.06 \times 1.5]) = \$100 \times 1.09 = \$109; \text{ but see below.}$$

How many days in a year?

A period of 92 days is not necessarily $^{92}/_{365}$ths of a non-leap year. Some treasurers use an ordinary interest year of 360 days. Others use an exact interest year of 365 days even in a leap year. The difference could be critical. $100m at 10% a year for $^{92}/_{360}$ths of a year is worth $41,895 more than the same amount over $^{92}/_{366}$ths of a year.

The bond market is one area where important variations in the length of a year exist, as the following illustrates.

- ◪ Standard practice in many European countries is to pretend that there are 12 months of 30 days each, making a 360-day year. The Euribor (European interbank offered rate) is calculated on this basis.
- ◪ Non-government bonds in the USA also accrue interest over a year of 360 days. However, while all months are treated as 30 days in length, one extra day is allowed if the interest period begins on the 30th or 31st of a month and ends on the 31st of the same or another month.
- ◪ Interest on US Treasuries (government bonds) is based on the actual number of days in the period divided by twice the number of days in the surrounding semi-annual period.
- ◪ British and Japanese bond interest is based on the actual number of days divided by 365 even in a leap year. But calculations relating to British gilts (government bonds) recognise February 29th, while those relating to Japanese bonds do not.
- ◪ Some Euro-sterling issues accept that there are 366 days in a leap year.

Simple returns and payback periods

The accountant's payback period and return on investment are based on the arithmetic of simple interest. If €1,000 spent on a stock market gamble or a capital project produces receipts of €100, €400, €500 and €1,600 over four years, the gross return is the total, €2,600. The net return is €1,600.

Return on investment. The simple percentage return is net return divided by outlay: 1,600 ÷ 1,000 = 1.60, or 160%. Such a figure is misleading. This total rate of return might have some value for quick comparisons with other projects, but among other things it ignores the pattern of income. If the receipts in this example were reversed, with the €1,600 coming in the first year, the overall rate of return would still be 160%, but time preferences and risk would be greatly improved.

Payback period. The payback period is the time taken to recover the original outlay, three years in this instance. If (estimated) returns were known in advance, the project might be shelved in favour of one with a two-year payback. The reasoning is based on the ideas of time preferences and time risk, but once again it ignores the pattern of receipts and what happens after the end of the payback period. The whopping €1,600 receipt ignored by payback analysis might make this project better than rival schemes. Payback is a useful concept, but it should not be used alone.

Payback and the simple rate of return are sometimes used together, but there are much better ways of looking at time and money, as will become clear.

Compound interest

If interest on a bank deposit is put into the account at the end of year 1, the principal sum for year 2 is greater by the amount of the interest. There is a larger sum now earning interest. Interest that itself earns interest is called compound interest. The total return depends on the frequency of compounding, the conversion period.

For example, €1m invested for 12 months at 10% a year generates €100,000 in interest if the interest is paid only once at the end of the year. With two conversion periods, the total interest is €102,500. Think of this as simple interest twice. For each half year, the interest accumulation factor is $1 + {}^{0.10}\!/\!_2 = 1.05$. After the first six months the principal is €1,000,000 × 1.05 = €1,050,000. This higher sum earns the same rate of interest for another six months: €1,050,000 × 1.05 = €1,102,500. The arithmetic could be written as €1m × 1.05 × 1.05. With four conversion periods, the interest accumulation factor is $1 + {}^{0.10}\!/\!_4$ or 1.025. Annual interest is €1m × 1.025 × 1.025 × 1.025 × 1.025 = €1,103,813.

Growth rates and exponents

Deal with growth rates over any period using proportions and "multiple multiplication". Growth of 5% a year ($r = 0.05$) due to interest or inflation increases an original sum by a factor of 1.05 ($1 + r$) after one year. After two years, the accumulation is 1.05×1.05. After three years, $1.05 \times 1.05 \times 1.05$ and so on. Multiplying a number x by itself y times is written x^y for short and known as raising x to the power of y. This y is sometimes known as an exponent. The accumulation factor for growth over n years is $(1 + r)^n$. For example, 100 growing by 5% a year for three years increases to $100 \times 1.05^3 = 115.8$. Basic rules for handling exponents are shown below.

General rule	*Example*
$x^y = x \times x \times x \times x \times x \dots \times x$ (y times)	$5^3 = 5 \times 5 \times 5 = 125$
	$-5^3 = -5 \times -5 \times -5 = -125$
	$0.5^3 = 0.5 \times 0.5 \times 0.5 = 0.125$
$x^0 = 1$ (where x is not zero)	$5^0 = 1 \quad -5^0 = 1 \quad 0.5^0 = 1$
$x^1 = x$	$5^1 = 5 \quad -5^1 = -5 \quad 0.5^1 = 0.5$
$x^{-y} = 1 \div x^y$ (where x is not zero)	$5^{-3} = 1 \div 125 \quad -5^{-3} = 1 \div -125$
$x^{1/y} = \sqrt[y]{x}$ (the yth root of x)	$125^{1/3} = 5 \quad 0.125^{1/3} = 0.5$
$x^y \times x^z = x^{y+z}$	$5^2 \times 5^3 = 5^{2+3} = 5^5 = 3{,}125$
$x^y \div x^z = x^{y-z}$	$5^3 \div 5^2 = 5^{3-2} = 5^1 = 5$

Two things happen with compound interest. First, the interest accumulation factor (1.10, 1.05, 1.025) shrinks as the frequency of compounding increases. Arithmetically, it is $1 + (i/k)$, where i is the rate of interest and k is the number of conversion periods a year. Second, the multiplier is multiplied by itself once for each conversion period; for example, $1.025 \times 1.025 \times 1.025 \times 1.025$, which can be written much more tidily as 1.025^4.

In math-speak, multiplying a number by itself n times is called raising it to the power of n. Your calculator's x^y key (the letters may be differ-

Table 2.1 **Critical compounding: the effect of 10% per annum interest compounded at different frequencies**

Interest compounded	No. of conversion periods a year	Value of $1m after 1 year ($m)	Equivalent rate with annual compounding (%)
Annually	1	1.1000	10.00
Six monthly	2	1.1025	10.25
Three monthly	4	1.1038	10.38
Monthly	12	1.1047	10.47
Daily	365	1.1052	10.52
At every moment in time	Infinite	1.1052	10.52

ent) reduces interest rate sums to a single sequence of button-pushing. For $2m earning 10% interest compounded every three months for a year, punch $2 \times 1.025\ x^y\ 4 =$ and the answer 2.207626 (ie, $2,207,626) pops up.

It is essential to grasp the idea of the interest accumulation factor. With 10% interest compounded once a year, a principal sum grows at a proportionate rate of 1.10 a year. After four years it has grown by $1.10 \times 1.10 \times 1.10 \times 1.10$. Thus, the interest accumulation factor is 1.10^4 or 1.4641.

Unravelling an interest accumulation factor is simple. Just as $1.10^4 =$ 1.4641, so $1.4641^{1/4} = 1.10$. Note that $x^{1/y}$ is read the yth root of x. Most calculators have a key for this operation. This $x^{1/y}$ is to x^y what $\div$ is to $\times$.

One basic assumption here is that the principal earns interest from the first day of the conversion period. If the principal is earning interest for, say, eight days in any conversion period, the interest factor for that period is perhaps $r \times 8/365$ (but see How many days in a year?, page 36).

Critical conversion periods
Clarity of thought and speech is essential when dealing with compound interest. Table 2.1 shows the value after one year of $1m earning 10% a year with different conversion periods. In every case, the interest is correctly quoted as 10% a year; with different compounding periods, total interest earnings vary by over $5,000. The message is clear. The conversion period, the frequency of compounding, is critical. To be told that the interest rate is, say, 10% a year should automatically prompt the question: "What is the conversion period?". "10% a year compounded every six months" is tight terminology.

Compound interest

General rule   *Example*

Where

PV = present value (principal) $1,000

i = annual rate of interest as proportion   0.125 (ie, 12.5%)

k = no. conversion periods per year   4 (quarterly interest)

n = no. conversion periods in total   8 (2 years)

Future value $FV = PV \times [1 + (i \div k)]^n$
$$= \$1,000 \times [1 + (0.125 \div 4)]^8$$
$$= \$1,000 \times 1.03125^8$$
$$= \$1,279.12$$

Compound interest $CI = FV - PV$
$$= \$1,279.12 - 1,000 = \$279.12$$

Effective interest rate
$$r_e = [1 + (i \div k)]^k - 1$$
$$= [1 + (0.125 \div 4)]^4 - 1$$
$$= 1.03125^4 - 1$$
$$= 1.1310 - 1$$
$$= 0.1310 \text{ (ie, 13.10%)}$$

Present value $PV = FV \div [1 + (i \div k)]^n$
$$= \$1,279.12 \div [1 + (0.125 \div 4)]^8$$
$$= \$1,279.12 \div 1.03125^8$$
$$= \$1,000$$

It is interesting to note that there is an upper limit on the return that can be achieved by increasing the frequency of compounding. If the interest rate is r, the once-a-year equivalent of compounding at each instant of time is $e^r - 1$, where e is the constant 2.71828. As indicated in Table 2.1, 10% compounded continuously is the same as a rate of $e^{0.10} - 1$ = 10.52% compounded once a year.

Effective interest rates

The best way of coping with interest rates with different conversion periods is to reduce them all to some standard period, as in the final column of Table 2.1. Any period would do, but it is logical to define the effective interest rate as the rate which would be earned (or paid) if

Discounting

General rule	*Example*
Where	
FV = redemption (repayment) value	$1,000
i = annual rate of interest as proportion	0.075 (ie, 7.5%)
n = time period in years	0.25 year (ie, 90 ÷ 360 days)
Present value PV = FV × [1 − (i × n)]	= $1,000 × [1 − (0.075 × 0.25)]
	= $1,000 × 0.98125
	= $981.25
Discount D = FV − PV	= $1,000 − 981.25 = $18.75
Effective interest rate	
$r_e = (i \div PV) \div n$	= ($18.75 ÷ 981.25) ÷ 0.25
	= 0.0191 ÷ 0.25
	= 0.0764 (ie, 7.64%)

interest was compounded just once a year.

Fortunately, this is straightforward. There is a simple relationship that links interest rates. If i is the annual rate of interest as a proportion and k is the number of conversion periods in one year, then the effective rate of interest r_e is found from this vital relationship: $(1 + r_e) = [1 + (i \div k)]^k$.

For example, with 10% a year compounded every three months (quarterly): $(1 + r_e) = [1 + {}^{0.10}/4]^4 = 1.025^4 = 1.1038$. So the effective equivalent is 10.38%.

Note the importance of terminology. For example, 10% a year compounded quarterly (2.5% a quarter) is not the same as quarterly compounding which yields 10% a year growth (2.41% a quarter).

The key compound interest formulae are outlined on page 40 in a general purpose format. From now on, all examples use an effective interest rate (once a year compounding) unless otherwise stated. Use the relationships illustrated to convert to and from other rates.

Discounting

Sometimes interest is charged in advance. For example, a lender making an advance of €10,000 at 10% might hand over proceeds of only €9,000 to the borrower, with €1,000 discounted as interest. The borrower would repay the full €10,000. A typical example is a money market bill. The borrower issues the paper at, say, €97 and redeems it perhaps 91 days later for its par (face) value of €100. The difference between the two amounts, the discount of €3, takes the place of an interest payment. (The arithmetic is shown on page 41.)

In discounting parlance, the rate of interest (discount rate) in the money market example is $3/100 = 0.03$ or 3% (pretend the term is 12 months). But since a present value of €97 attracts interest of €3, the interest rate is effectively $3/97 = 3.1\%$.

An even bigger discrepancy occurs with discounting of interest on many personal or consumer loans. For example, an advance of $1,000 at 12.5% discounted in advance and repayable over 12 months attracts interest of $125. The principal is grossed up and repaid in instalments of $1,125 \div 12 = \$93.75$. But since the balance declines steadily the effective interest rate is actually 24.75%. This is nearly double the quoted discount rate.

Interest rate comparisons

It should be clear that care is required when comparing interest rates. The trick is to convert all interest rates to effective rates using simple relationships (see Compound interest box, page 40). These reveal, for example, that 10% interest compounded four times a year yields an effective rate of 10.38% and is therefore a better deal than 10.25% interest added only once a year.

Legislation in many countries attempts to protect the consumer from sleight of hand such as discounting personal loan interest in advance. The effective rate must be quoted on all contracts in the USA. The UK goes a stage further, often requiring the disclosure of an APR, or annual percentage rate, which also includes charges for the loan, such as documentation fees or maintenance charges. These are interest in disguise.

Do not forget to include fees, charges, taxes and so on in your comparisons. For example, a $1,000 loan with interest of $100 and a $10 commission charge has an effective interest charge of $110 (or 11%). Similarly, a $1,000 investment which produces a return of $100 from which is deducted $10 withholding tax is effectively yielding $90, or 9%. It is especially important to take such "bonuses" into account when comparing loans or investments with differing hidden extras.

Annuities
Streams of payments
Comparing interest rates is only part of the story. Interest rate problems in general always involve the interest accumulation factor and any two of present value, future value and a stream of payments.

Streams of regular payments (and therefore receipts) at equal time intervals are called annuities. Most streams may be thought of in terms of ordinary annuities, where the payment (known as the rent or payment size) is made at the end of the payment interval.

Typical examples include: a lump sum invested with an insurance company to be paid back in ten annual instalments (the annuity) when the investor retires; a series of payments (the annuity) to pay off a debt such as a home mortgage; and regular transfers (the annuity) to a sinking fund to cover some future obligation, such as replacement of capital equipment. These are finite annuities with a fixed number of payments.

Perpetual annuities
A perpetual annuity, or perpetuity, goes on forever. A good example is a fund set up to pay a fixed scholarship once a year for ever more. At first glance, this might seem to involve a huge principal sum since the payments from it are never ending. But as soon as the payment is

---

2.1

Annuities in action

The regular savings described in the text are transferred to the scheme at the end of each year. What if the savings were made at the beginning of the year? Since the end of one period is the beginning of the next, the savings scheme illustrated would effectively last for four years. Exactly the same arithmetic would apply, but for one period longer. In annuity-speak, it would be an annuity due rather than an ordinary annuity. The future value of an annuity due is the same as that for an ordinary annuity multiplied by $(1 + r)$.

recognised as interest, it becomes easily manageable in size. An annuity paying €3,000 a year in perpetuity has a present value (ie, requires an investment of) €100,000 if the interest rate is 3% a year. The present value = R ÷ i where R is the regular payment and i is the interest rate. Government bonds with no repayment date (the coupon, or interest, is the annuity) are everyday examples of perpetual annuities.

The future value of a stream of payments

Finding the future value of a stream of payments is straightforward when approached through a single payment example.

A single payment. Suppose you put €100 in a fixed deposit for five years at an effective compound interest rate of 6% a year. The compound interest formula (page 40) indicates that the future value $FV = PV \times (1 + r)^n$ or, in numbers, $FV = €100 \times 1.06^5 = €133.82$.

A stream of payments. Consider now saving €100 a year for three years at an effective interest rate of 6% a year. What will be the maturity value (ie, future value) at the end of three years? This reduces to three bits of fixed principal arithmetic.

Suppose the first €100 is transferred to the saving scheme at the end of year 1. It earns compound interest for the remaining two years, accumulating to $€100 \times 1.06^2$. The second amount accumulates to $€100 \times 1.06^1$. The third €100 is not in the scheme long enough to earn any interest. In math-speak, it accumulates to $€100 \times 1.06^0$, where any number raised to the power $0 = 1$ (see Growth rates and exponents box, page 38). This can be written out as:

$$FV = (€100 \times 1.06^2) + (€100 \times 1.06^1) + (€100 \times 1.06^0)$$
$$= (€100 \times 1.1236) + (€100 \times 1.06) + (€100 \times 1)$$
$$= €112.36 + €106.00 + €100.00 = €318.36$$

So the savings would grow to €318.36. Figure 2.1 shows how to cope if the savings are transferred into the scheme at the start rather than the end of the year.

The arithmetic could be written out in general purpose symbols as $FV = [R \times (1 + r)^{n-1}] + [R \times (1 + r)^{n-2}] + ... + [R \times (1 + r)^0]$ where R is the regular payment, r is the interest rate, and n is the number of periods. With a little algebraic manipulation, the formula shortens to $FV = R \times [(1 + r)^n - 1] ÷ r$. This is a wonderfully simple piece of arithmetic. You can use it to find the future value of €100 saved each year with a 6% interest rate: $FV = €100 \times [1.06^3 - 1] ÷ 0.06 = €100 \times 3.1836 = €318.36$.

The present value of a stream of payments

As with future value, present value problems relating to streams are no more than several single payment problems lumped together.

A single payment. Suppose Mr Jones decides that he needs €10,000 in five years to pay for his daughter's wedding. How much should he put on deposit today? (What is the present value of the future sum?) The basic relationship is $PV = FV \div (1 + r)^n$ or €10,000 $\div$ 1.06^5 = €7,473.

In passing, note that $z \div x^y$ is the same as $z \times x^{-y}$. This can be used to rewrite the present value relationship as $PV = FV \times (1 + r)^{-n}$, which simplifies other stream of payments arithmetic.

A stream of payments. Suppose Mr Jones wants to provide for his other daughter's school fees. If he can invest at 12% a year and the fees are fixed at €10,000 a year for nine years, how much should be set aside to cover the obligation? It is not too hard to see that he needs to know the present value of nine separate fixed deposits. In general,

$$PV = [R \times (1 + r)^{-1}] + [R \times (1 + r)^{-2}] + \dots + [R \times (1 + r)^{-n}]$$

where, as before, R is the regular payment, r is the interest rate and n is the number of periods. Again, this shortens neatly, to

$$PV = R \times [1 - (1 + r)^{-n}] \div r$$

which is another simple yet powerful relationship.

The present value of the school fees is €10,000 $\times$ $[1 - 1.12^{-9}] \div 0.12$ = €10,000 $\times$ 0.6394 $\div$ 0.12 = €53,282.50. If Mr Jones puts €53,282.50 aside at 12% a year compound, it will produce a stream of payments of €10,000 a year for nine years.

Regular payments 1: repaying a loan

When dealing with loans in which interest and principal are amortised (repaid, or killed off) in equal monthly instalments, remember that lending and borrowing are two sides of the same coin. Consider a home mortgage of $100,000 repayable over 20 years with 10% a year interest compounded monthly. What is the monthly repayment? From the lender's viewpoint, this is an investment with a present value of $100,000 which will produce an income stream for 240 months.

The present value and rate of interest are known, the regular repayment R is to be found. The present value formula can be rearranged as $R = PV \div [\{1 - (1 + r)^{-n}\} \div r]$. Note that this time, with 12 conversion periods a year, $r = 0.10/12 = 0.00833$. Slotting in the numbers: $R = \$100,000 \div [1 - (1 + 0.00833)^{-240}] \div 0.00833 = \$100,000 \div 103.62 = \$965.06$.

So the monthly repayment is \$965.06. The accumulation factor 103.62 remains fixed whatever the amount of the loan. If the loan was increased to \$500,000, everything else remaining unchanged, the monthly repayments would be \$500,000 ÷ 103.62 = \$4,825.11.

Regular payments 2: sinking funds

Mr Todd, the teddy bear manufacturer, realises that he will have to replace his stuffing machine in five years' time. He anticipates spending \$20,000. How much should he deposit quarterly in a sinking fund earning 8.5% a year compounded quarterly?

This time the future value and interest rate are known, so the repayment R can be found by rearranging the future value formula encountered earlier. As with the loan repayment example, note that $r = 0.085/4 = 0.02125$ and $n = 20$ periods. $R = FV \div [\{(1 + r)^n - 1\} \div r]$ gives $R = \$20,000 \div [\{1.02125^{20} - 1\} \div 0.02125] = \$20,000 \div 24.602 = \$812.94$. Mr Todd should put \$812.94 into his sinking fund each quarter to pay for the new machine in five years' time.

Streams and interest rates

The final question which crops up with stream-of-interest problems arises when interest rates are unknown. Suppose Mr Jones receives an inheritance of €53,282.50. If he wants to convert it into a stream of payments of €10,000 a year to cover nine years' school fees, what minimum interest rate would he require?

He knows the regular payment (R) and the present value (PV); he needs to know the interest rate (r). The starting point must be the formula bringing these together: $PV = R \times [1 - (1 + r)^{-n}] \div r$. This will not rearrange very helpfully. For two periods, when $n = 2$, there is an easy solution. For three or four periods, the arithmetic becomes very complex. Beyond this, the answer is found only by trial and error or from tables of solutions to common problems.

The trial and error approach is actually quite straightforward. Pick a rate, say 10%, and use it to calculate the present value: €57,590.24. A higher present value is required. Pick another rate, say 14%, and try again: $PV = €49,463.72$. These two figures straddle the required rate, so it must be somewhere in between 10% and 14%. If Mr Jones picks 12% next, he will be spot on.

Many interest rate problems can be solved using published tables covering a range of common investment situations. However, they are becoming increasingly obsolescent. PC spreadsheets or financial calculators provide routines which perform the trial and error calculations to identify interest rates in compound interest problems.

Investment analysis

Competing investment opportunities are easily reviewed using a simple application of compound interest arithmetic. There are two main approaches, net present value and internal rate of return.

Key financial formulae

This is the most important box in this chapter. It provides the key to almost any money problem. Just identify which variables are known and which one is to be calculated, and choose the appropriate formula.

Where   PV = present value

FV = future value

R = stream of income per conversion period

n = number of conversion periods in total

r = rate of interest per conversion period as a proportion

k = number of conversion periods per year

If $k > 1$ then $r = i / k$ (where i is the annual compound rate of interest)

Single payment

$$PV = FV \div (1 + r)^n$$
$$FV = PV \, (1 + r)^{-n}$$

Infinite stream of payments

$$PV = R \div r$$

Finite stream of payments

Formulae linking PV, R, r and n:

$$PV = [R_1 \times (1 + r)^{-1}] + [R_2 \times (1 + r)^{-2}] + \ldots + [R_n \times (1 + r)^{-n}]$$
$$PV = R \times [\{1 - (1 + r)^{-n}\} \div r]$$

Formulae linking FV, R, r and n:

$$FV = [R_1 \times (1 + r)^{n-1}] + [R_2 \times (1 + r)^{n-2}] + \ldots + [R_n \times (1 + r)^0]$$
$$FV = R \times [\{(1 + r)^n - 1\} \div r]$$

Note. Net present value = PV − outlay.

Net present value

Net present value is simply present value less outlay. Remember that present value is the amount of money that would have to be invested today to produce a future sum or stream of income.

For example, a machine is expected to produce output valued at €1,250 next year, €950 in year 2, €700 in year 3 and €400 in year 4. What is the present value of this stream of income, assuming an interest rate of 10% a year?

This involves a finite stream of income where the known facts are the rate of interest (r), the number of periods (n), and the stream of income (R$_1$, R$_2$, ... R$_n$). The present value is to be found.

Using Key financial formulae (see box, page 47), the appropriate formula is:

$$PV = [R_1 \times (1 + r)^{-1}] + [R_2 \times (1 + r)^{-2}] + ... + [R_n \times (1 + r)^{-n}]$$
$$PV = (€1{,}250 \times 1.10^{-1}) + (€950 \times 1.10^{-2}) + (€700 \times 1.10^{-3}) + (€400 \times 1.10^{-4})$$
$$= €2{,}720.61$$

The present value of the machine's output is €2,720 (approx.). If the machine cost €1,000, its net present value is €2,720 − €1,000 = €1,720.

A positive net present value, as here, indicates that a project earns more than the chosen interest rate, or rate of return (10% in this example). If a project produces a negative net present value, its true rate of return is below the chosen discount rate.

If there are several investment projects (financial or capital), they can be compared using net present value. Quite simply, the project with the highest net present value is the one to opt for, but note that the best option at one discount rate may not be the optimal project at another rate. It is worth trying a what-if approach using several interest rates. Choice of the most appropriate rate is discussed below. Spreadsheets

Table 2.2 **Comparing internal rates of return**

	Project A	Project B	Marginal analysis Project C
Outlay ($)	1,000	2,000	1,000
Gross return ($)	1,200	2,300	1,100
Yield (%)	20	15	10

and financial calculators usually have NPV functions. Typically, enter a stream of payments and a discount rate, and read off the net present value. Outlays are entered as negative amounts.

Internal rate of return

For investment appraisal the main alternative to net present value is internal rate of return, also known as discounted cash flow or yield. This is the compound interest rate which equates the present and future values.

Take an example. A machine costing €1,000 will produce output of €1,250 next year, €950 in year 2, €700 in year 3 and €400 in year 4. What is the yield of this investment? Present value (−€1,000), the payment size (R), and the number of payments (n) are known, the rate of interest (r) is to be found. Again the appropriate formula is $PV = [R_1 \times (1 + r)^{-1}] + [R_2 \times (1 + r)^{-2}] + ... + [R_n \times (1 + r)^{-n}]$. This time, trial and error or an electronic friend produce the answer 97%.

Spreadsheets or financial calculators are invaluable. As with NPV, outlays are treated as negative income. For example, with a spreadsheet −1,000, 1,250, 950, 700 and 400 would be entered in a continuous range of cells and the IRR function keyed into another, where the answer would appear.

Internal rate of return is an excellent measure with a good intuitive feel. Percentages are familiar. They make it easy to compare competing investments and the cost of capital. Simply line up the IRRs and see which is highest.

A problem with the internal rate of return is that the equations can produce more than one yield (when negative cash flows generate more than one positive root). In other words, if there is more than one change of sign it might make IRR unworkable. This tends to happen with capital projects rather than with portfolio analysis where negative cash flows are somewhat unwelcome.

Marginal returns

Internal rate of return has one defect. It does not reveal anything about capital outlays. What if two projects have the same yield, but one requires twice the expenditure of the other? Where the outlays are different, extend the usefulness of internal rate of return analysis by comparing marginal returns.

For simplicity, consider two 12-month investments shown in Table 2.2. Project A yields 20% for a $1,000 outlay. Project B returns 15% for a

$2,000 investment. Project A is the clear winner. But assume for a moment that the pattern of expenditure and receipts is more complex and the answer is not obvious. Project C is the difference between Projects A and B; it is the marginal outlay and return of Project B compared with Project A. For each period, C = B − A. The 10% yield on this third project must be calculated separately; subtraction does not work.

Project B can be thought of as two separate projects: one with exactly the same outlays and returns as A; and one with the 10% return revealed in column C. In this clear cut case, B might be rejected because its marginal return of 10% on the additional outlay (over that for Project A) is perhaps below the cost of capital. In more testing decision-making, such marginal analysis is highly revealing.

Investment appraisal and discount rates
What interest rate should be used for investment appraisal (whether through net present value or internal rate of return)? For financial investment, any project should be better than a safe alternative, such as government bonds or a bank deposit. For capital investment, the project might also have to outrank the cost of capital or the overall rate of return on capital employed (profit as a percentage of capital). Obviously, a return below any of these is a loser in pure cash terms.

Using net present value and internal rate of return
Net present value and internal rate of return are the same thing approached from different angles. The internal rate of return is perhaps slightly more complicated and sometimes even impossible to calculate. But it is generally preferable since it cuts straight to the bottom line.

The payback period can be brought into the analysis now, and another useful concept is the profitability index: the net present value of net earnings divided by the present value of total capital outlays. The higher the result of this sum, the better the profitability of the investment project. It goes without saying that comparisons of investments should include factors which vary between projects: taxes, fees, depreciation, and so on.

If two projects cover the same term and if a constant rate of inflation is projected, there is no need to adjust for inflation in order to choose between them. A ranking of their relative merits is unaffected by a constant inflation factor. Where terms and inflation rates vary inflation should be taken into account.

Inflation

Inflation throws an irritating spanner into the works. Fortunately, accounting for inflation is easy in a numerical sense. The ugly dragon is nothing more than an interest charge. The annual rate of inflation is compounded yearly.

If inflation is projected at 10% a year, how much should Mr Jones leave in a non-interest bearing account to have €10,000 of current spending power five years hence? Inflation of 10% over five years produces compound growth of $1.10^5 = 1.6105$; just over 61%. This 1.10^5 is the inflation accumulation factor. It is exactly the same conceptually and arithmetically as the interest accumulation factor. So €10,000 of spending power today equates to €10,000 × 1.6105 = €16,105 in five years, assuming the 10% inflation forecast is correct.

But what if the deposit earns interest over the five years? Quite simply, all Mr Jones has to do is put away a present value that generates a future value of €16,105. For example, with 6% a year interest rates, $16,105 × 1.06^{-5} = €12,034$. This is larger than €10,000 because inflation of 10% is outstripping interest of 6%.

Another way of looking at the problem is to note that each year the spending power of the money increases by 6% interest then decreases by 10% inflation. The net annual rate of change is 1 − (1.10 ÷ 1.06) = 1 − 1.0377 = −0.0377, which implies that the spending power of money shrinks by 3.8% a year. (The division reflects the fact that interest and inflation are moving in opposite directions; interest is adding to spending power, inflation is eroding it.) Do not derive an aggregate rate of inflation plus interest by addition or subtraction because this does not take account of compounding. Over five years the accumulation factor is $1.0377^5 = 1.2034$. So €10,000 at a discount rate of 3.8% is €10,000 × 1.2034 = €12,034; the same answer as before.

Sometimes a future value is known and it is useful to establish its current value given some rate of inflation. Say, for example, in two years you will receive $10,000. If inflation is projected at 8% this year and 6% in the second year, what is the current value of this sum? The inflation accumulation factor is 1.08 × 1.06 = 1.1448 (ie, cumulative inflation is 14.48%.) The current value of the sum is $10,000 ÷ 1.1448 = $8,735. If interest rates are 5% a year, the present value of this current value is $8,735 ÷ 1.1025 = $7,923.

There are two important points to note. First, the effects of inflation and interest must be compounded. Do not overlook the way that they can move in opposite directions. Second, remember that inflation forecasts are subject to error. Also, published inflation rates apply to a basket of goods and services. A specific project for which a sum of money is set aside may suffer a totally different rate of inflation from the basket.

For complex investment projects, specify carefully the assumed inflation rates and convert each future value into a current value, then perform the internal rate of return or net present value calculations on these current values.

Interest rate problems in disguise

Many business decisions are interest rate problems in disguise. The trick is to undertake a careful analysis, when the solution becomes obvious.

Lease and rental charges

Problems involving leases and rentals are not difficult to solve. In effect, the lessor is lending a sum of money to the lessee and recovering it in instalments. It just happens that the loan is used to buy capital equipment. This is a present value, interest accumulation and size of payments problem. Two brief examples will suffice.

How much to charge. Capital Computers (CC) buys computer printers for €500 each and leases them to business users. What quarterly rental should CC charge to recover the expenditure within two years and generate a 15% annual return? The return needs to be specified carefully. In this case, the problem might be rephrased as: "If CC puts €500 on deposit at 15% a year compounded quarterly and withdraws it over eight quarters, what is the size of the regular withdrawal?"

The relationship is $R = PV \div [\{1 - (1 + r)^{-n}\} \div r]$ (see Key financial formulae box, page 47), where the rate of interest $r = 0.15/4 = 0.0375$. So $R = €500 \div [1 - 1.0375^{-8}] \div 0.0375 = €73.50$. This is CC's desired rental charge.

Rent or buy? Should Yvette rent one of these printers at €73.50 a quarter or buy at the retail price of €700? She believes that with technological change the life of the printer is only 18 months. She currently has the money on deposit at 9% a year compounded quarterly. One approach is to find the stream of payments that the deposit would produce over eight quarters. Note that $r = 0.09/4 = 0.0225$. From the same relationship used to set the rental charge, $R = PV \div [\{1 - (1 + r)^{-n}\} \div r]$, we have $R = €700 \div [(1 - 1.0225^{-8}) \div 0.0225] = €96.59$. This is more than €73.50. She should leave her money in the bank, draw it down to cover the rental charge, and pocket the surplus. An alternative would be to compare the cost of renting with the cost of repaying a loan taken out to buy a printer.

Note that these examples have conveniently ignored taxes, maintenance, other costs such as air time and the residual value of the equipment. Allowance must be made for such factors in real life by adjusting the payments up or down. Residual value would be factored in as the last term of the expanded version of the equation used in this example:

$$PV = [R_1 \times (1 + r)^{-1}] + [R_2 \times (1 + r)^{-2}] + ... + [R_n \times (1 + r)^{-n}]$$

Discount for cash?

What do you do if you are offered a discount for immediate payment? Convert the discount into an annual percentage rate and see if it is better than the interest you earn on your deposit (or less than the interest you pay on your loan).

For example, a supplier proposing full payment within 31 days or 1% discount for immediate settlement is offering you a flat 1% interest payment for 31 days' use of your money. This grosses up to 12% a year (remember that it is not compound). If your cash is on deposit earning less than 12% a year, grab the discount. If you have to borrow the money, the discount is worthwhile if the loan costs less than 12% a year. In effect, you could be borrowing from your bank at 10% and lending to your supplier for 12%. Thus you would make a 2% annual return.

An offer in cash terms, such as full payment of €2,000 within 90 days or €50 discount for immediate payment, should be converted into a percentage rate. This one is $50/2000 = 2.5\%$ for three months, or 10% a year. As before, accept the discount if it is better than whatever else you would do with the money.

Similarly, if you are the seller, offer a discount which you can finance without making a loss. Either way, do not forget to include administration costs, taxes, and so on in the calculations.

Exchange rates

Exchange rates are simply the price of one unit of money in terms of another currency. Two things cause difficulties: one is terminology; the other is the period for which a rate applies.

Exchange rate terminology and arithmetic

It is not very helpful to say that the exchange rate between the dollar and the euro is 1.10, although money changers do this all the time. Do you receive €1.10 for a dollar or $1.10 for a euro? The rate is €1.10

reveals the answer immediately, but the currency is rarely quoted with the number. Take a look at the financial pages of any newspaper. Loose but generally accepted terminology is: the exchange rate of the euro against the dollar is 1.10. Better is: the euro per dollar rate is 1.10. This makes it clear that the rate is denominated in terms of the dollar. In every case, the message is $1 = €1.10, which is the tightest terminology of all.

The important thing to do is to identify the currency which goes into the equation with a value of 1. When the exchange rate increases, this currency is getting stronger or appreciating. When it shrinks, the currency is depreciating. In other words, $1 = €1.20 indicates that the dollar is stronger than at a rate of, say, $1 = €1.10. One dollar will buy more European goods and services at the higher rate. Of course, the converse is also true. The other currency weakens as the first one strengthens. The European currency is weaker at $1 = €1.20 than at $1 = €1.10.

Some big statistical blunders are often made when calculating the percentage appreciation or depreciation of a currency. The tables in the back of The Economist each week show the latest exchange rate for the dollar against various currencies and its rates a year ago. Suppose that last year $1 = €1.00, compared with $1 = €2.00 today. This means that the dollar has appreciated by 100%, ie, (2.0 − 1.0) ÷ 1.0 × 100%. A common mistake, however, is to say that the euro has depreciated by 100%. This is clearly nonsense: a 100% fall would imply that a currency is now worthless. The trick in calculating the size of the euro's depreciation (as opposed to the dollar's appreciation) is first to re-express the exchange rate in terms of

Table 2.3 **Exchange rates and time**

Date	Profits remitted €m	Exchange rate €1 = $	$ equivalent
Conversion using actual exchange rates			
March 31	18.25	0.79	14.42
August 25	25.75	1.17	30.13
Total transactions	44	1.01	44.55
Conversion of total using other exchange rates			
January 1	44	0.81	35.64
December 31	44	1.19	52.36
Annual average of daily rates	44	1.15	50.60

the euro, ie, the euro has fallen from €1 = $1.00 to €1 = $0.50 over the past year. This represents a euro depreciation of 50%.

There are two tricks you can perform with exchange rates.

- Swap base currencies by dividing the rate into 1. If €0.91 = $1, then the dollar per euro rate is €1 = $¹⁄₀.₉₁ = $1.10.
- When you know two rates, you can calculate a third. If €1 = $1.05, and €1 = SFr1.47, then $1.05 buys SFr1.47; or $1 = 1.47/1.05 = SFr1.40.

Talking to dealers. Foreign exchange dealers usually quote two-way exchange rates, such as a dollar per euro rate of, say, 1.0565–1.0575. This means that they buy the dollar at €1.0565 per $1 (the bid rate) and sell the dollar at €1.0575 per $1 (the offer rate). Common sense will show which is the bid-to-buy and which is the offer-to-sell rate, since the dealers will not set themselves up to make a loss. A spread as narrow as the one shown here (1.0565 – 1.0575 = 0.10 cents) would apply to an inter-bank deal for perhaps £5m. A much wider spread would be quoted for smaller amounts. Exchange rates for immediate settlement are known as spot rates.

Exchange rates and time

An exchange rate is the price of foreign money at one moment in time. Always compare like with like. The top half of Table 2.3 shows profits remitted by a foreign subsidiary to its parent company. Using exchange rates ruling on the dates of the two transactions, the total of €44m was converted into $44.5m. However, the devious parent company lumps together all the transactions during the year and shows them as one figure in its annual report. The lower half of Table 2.3 shows how the parent company can create three different dollar amounts simply by using different reported exchange rates. Few commentators would object to using an average exchange rate, but look how it understates the dollar value of foreign profits.

Moreover, the average exchange rate in the final line of Table 2.3 was obtained by adding daily rates and dividing by the number of days in the year. An average of end-month rates would produce yet another exchange rate for the parent company to play with.

The message is always to check the basis for currency conversions.

Forward exchange rates

Forward exchange rates are used to determine today an exchange rate that will be applied on a given date in the future. For example, in one month's time an Italian manufacturer will be paid $1m on delivery of machine tools. The manufacturer will have to convert the dollars into euros to meet other commitments. It can lock in to a fixed exchange rate now by agreeing a forward rate with its bank. The manufacturer is then guaranteed a fixed amount of euros whatever happens to exchange rates in the meantime.

In essence, the bank borrows dollars for one month, sells them immediately for euros, and places the euros in the money markets to earn interest for one month. At the end of the month, the bank gives the euros to the manufacturer in exchange for its dollars, and uses the dollars to repay the dollar loan.

Thus, forward rate agreements are based not on explicit expectations of exchange rate movements, but on the spot (current) exchange rate and relative interest rates. For the bank, the cost of the transaction in this example is the interest paid on the loan in dollars less the interest received on the sterling deposit. The dollar's forward rate would be more expensive than the spot rate (at a forward premium) if dollar interest rates were below euro interest rates, or less expensive than the spot rate (at a forward discount) if dollar rates were above sterling rates.

Forward exchange rates are usually quoted as margins, or differences, from the spot rate. For example, if the spot dollar per euro rate is 1.0565–1.0575, the one-month forward rate might be quoted as a dollar discount or premium of 0.0113–0.0111 (ie, 1.13 to 1.11 cents or 113 to 111 points). The forward rates are therefore $1.0565 − 0.0113 = $1.0452 bid (bank buys the dollar) and $1.0575 − 0.0111 = $1.0464 offer (bank sells the dollar).

3 Descriptive measures for interpretation and analysis

"He uses statistics like a drunken man uses a lamp post, more for support than illumination."

Andrew Lang

Summary

Descriptive or summary measures such as totals, averages and percentages are powerful tools. They provide a concise shorthand way of describing any collection of related numbers, bringing order to an unruly mass of data. They also give access to some stunning techniques for analysis.

There are three key measures:

- ◪ the mean, which is an average, an everyday guide to the midpoint in a set of numbers;
- ◪ the standard deviation, which is a measure of how widely the numbers are spread around the mean; and
- ◪ standard distributions which describe the shape (pattern) in a set of numbers. The most common standard distribution is known as the normal distribution.

If you know that a set of numbers fits a standard distribution, and you know the mean and standard deviation, you know just about everything there is to know.

Distributions

Any set of numbers that can be summarised is a distribution: the distribution of salaries in a firm, the distribution of sales by branch, the distribution of house prices in Hungary. Whenever you come across an average, or any other summary measure, it is describing a distribution. To interpret the data properly, you need to be able to visualise the distribution.

For example, imagine that the salaries paid to each person in a large company are written on bricks; one brick per employee. If the bricks are

Summarising a distribution

3.1

Fat Cat Treats: number of packets purchased per household in one month

This graph shows the results of a survey by the distributors of Fat Cat Treats. Purchasers were asked how many packets their household had bought in the past month. The vertical bars record the results. For example, 48 households purchased just one packet, 85 households bought two packets, and so on. The bars together form a histogram. This is a special bar chart where the bars touch and the areas of the bars are significant. The line joining the tops of the bars is a polygon (Latin for many angles), which is carefully drawn so that there is the same area under it as there is in all the bars.

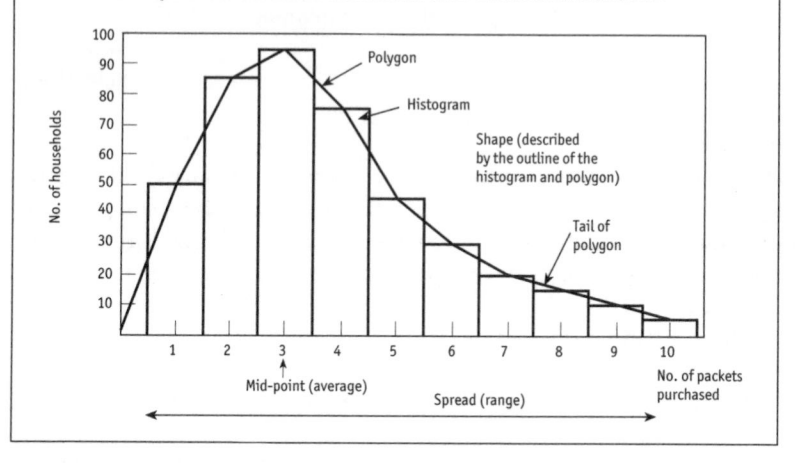

tipped into a random pile, the information on them is a meaningless jumble. You probably come across information in this form from time to time. Now, suppose that the bricks are lined up in order of the amounts written on them, with the lower values to the left. Bricks bearing the same amounts would be stacked on top of each other. Together the bricks would build a wall which would probably taper away at the ends and reach its maximum height a little way in from the left (not unlike the vertical bars in Figure 3.1). The wall represents the distribution of salaries paid by the company in question. It can be described with just three measures.

1 **An average.** Think of this as the amount written on the bricks in the tallest pile. If you knock the whole wall down and pick a brick at random it will most likely bear a value close to the average salary.
2 **A measure of spread.** This indicates the length of the wall. If you knock the wall down and pick a brick from the rubble, the amount written on it could not be outside the range of salaries actually illustrated by the wall.

3 **A measure of shape.** This reveals whether the bricks are placed symmetrically around the average, or perhaps skewed to one side. That is, whether the average is in the middle or off-centre and if off-centre by how much.

If you have never seen the *salaries wall* but someone gives you three pieces of information (average, spread and shape), you can immediately visualise what it looks like. These three measures bring order to a mass of data. You could use them as a framework from which to sketch the wall. From this sketch you could estimate where in the wall you would find any brick that happened to interest you, without having to go back to the raw data. This is especially useful if there are thousands of figures involved.

The three summary measures are also useful for comparing two or more sets of data (eg, salaries paid by two firms). In addition, they provide access to powerful analytical tools.

Averages

The average is the most commonly used summary measure. It identifies the mid-point of a distribution, and indicates where the distribution is located in the overall set of numbers from minus infinity to plus infinity.

Most of us think deeply about our own weight, fitness, brains, wealth

Table 3.1 **Salaries at Backstreet Byproducts**

	$		
1 Junior	15,000		
2 Clerk	21,600		
3 Secretary	23,400		
4 Production worker	28,800		← Mode
5 Production worker	28,800	← Median	← Mode
6 Production worker	28,800		← Mode
7 Production supervisor	32,400		
8 Office manager	36,000		
9 Managing director	360,000		
TOTAL	574,800 ÷ 9 = $63,867		
	↑		
	Mean		

and success relative to the average. In fact, there are many ways of calculating an average. This is helpful when we do not compare well using one approach.

Mean. The most familiar average is obtained by adding all the numbers together and dividing by the number of observations. This gives a single figure which is a sort of balancing point or centre of gravity for a distribution. For this reason, it is known as an arithmetic mean, or more loosely, a mean (from the Latin for middle).

The mean is the most versatile average. It is widely understood, easy to calculate, and it uses every value in a series. However, there are drawbacks. Consider the salaries listed in Table 3.1. The mean salary is $574,800 ÷ 9 = $63,867, but just about everyone at Backstreet Byproducts earns less than half this amount; only the managing director takes home more than the average.

Median. An average which is not affected by extreme values (outliers) is a useful alternative to the mean. One such measure is the median. This is simply the value occupying the middle position in a ranked series. Put the observations in order and read off the central amount. In Table 3.1, there are nine salaries so the median, the fifth, is $28,800. With an even number of observations, add the two central values and divide by 2. The median of the series 4, 5, 8, 10 is (5 + 8) ÷ 2 = 6.5.

The median salary is a useful concept, since it is an amount which splits a group into two halves: 50% of employees earn less than the

---

Other averages

Two other important averages deserve mention.

Weighted average. The weighted average (weighted mean) was introduced on page 15. The components are each assigned a weight to reflect their relative importance. Many exchange rate indicators and stockmarket indices, such as sterling's trade-weighted index and the Dow Jones index, are weighted averages.

Geometric mean. The geometric mean is used to average a sequence of numbers which are growing. To find the geometric average of 12 monthly values, multiply them together and take the 12th root.

The geometric mean of 100, 110, and 121 is $\sqrt[3]{(100 \times 110 \times 121)} = \sqrt[3]{1,331,000} = 1,331,000^{1/3} = 110$.

median wage, 50% earn more. Clearly, it is more realistic for a Backstreet worker to aspire to the median rather than the mean wage.

Mode. The third main average, the mode, is the most fashionable value in a series. This is the average that Mr Day is thinking of when he tells you that on average he sells more 1 litre cans of paint than any other size. The most common salary, the mode, in Table 3.1 is $28,800. Not all series have a single mode. The following sequence does not: 1, 2, 2, 4, 4, 6, 7, 8, 9. It is bimodal (modes = 2 & 4); some series are multimodal.

Choosing and using the average. As a general rule, prefer the mean unless there is a good reason for selecting another average. Use the median or mode only if they highlight the point you want to make, or if the data cannot have a mean.

- ◪ Categorical data (eg, sales/marketing/personnel) can be summarised only with a mode. You can find which is the most popular (modal) department, but you cannot arrange the departments in numerical order or perform calculations such as (sales + marketing) ÷ 2.
- ◪ Ordinal data (eg, poor/fair/good) have both a mode and a median. You can find the most frequent result (mode) and rank them in order to find the median, but again arithmetic calculations are impossible (ie, poor + fair + good is meaningless).
- ◪ Interval and ratio data (1/2/3) can have a mode, median and mean since arithmetic calculations are possible with such data. In this case (1 + 2 + 3) ÷ 3 is meaningful.

Avoid averaging two averages. For example, the mean of (2, 4, 6) is 4 and the mean of (3, 5, 7, 9) is 6. The mean of 4 and 6 is 5. But the mean of the two sets combined (2, 4, 6, 3, 5, 7, 9) is 5.14. With real life situations the discrepancies are frequently much larger. Either go back to the raw data or apply a little cunning. If you know that three numbers have a mean of 4, they must sum to 12 (12 ÷ 3 = 4). Similarly, four numbers with a mean of 6 must total 24 (24 ÷ 4 = 6). The two totals, 12 and 24, add together to reveal the sum of the combined data. Since there are seven numbers in total, the mean of the combined series is (12 + 24) ÷ 7 = 5.14. This was deduced without knowledge of the actual numbers in the two averages.

Lastly, if presented with an average, always ask on what basis was it calculated.

Carving up a distribution

Just as a median splits a distribution into two halves, so there are some other useful dividing points.

Quartiles divide a series into four parts, each containing an equal number of observations. To identify the quartiles, simply rank the numbers in order and identify the values which split them into four equal segments. In the following illustration, the first, second and third quartiles are 4, 5 and 6.5 respectively. The second quartile is better known as the median.

Set of numbers → 2, 3, 3, 4,   4, 4, 5, 5,   5, 6, 6, 6,   7, 7, 7, 8

                          ↑   ↑   ↑

Quartile values   4   5   6.5

Deciles divide a series into ten equal parts. When you boast that your company is in the top 10%, you are saying that it is in the tenth (top) decile.

Percentiles split a series into 100 equal parts.

Using -iles. Strictly speaking, quartiles, deciles and percentiles are the values at the dividing points. In the illustration here, the first quartile value is 4. Confusingly, the terms are also used loosely to refer to the observations enclosed between two -iles (in this example, 2, 3, 3 and 4 would sometimes be called the first quartile observations).

Clearly it is difficult to carve up a distribution if there is not a conveniently divisible number of observations, and impossible if there are not enough.

Spread

An average becomes much more meaningful when it is accompanied by an indicator of spread (also known as scatter and dispersion). If someone tells you that the average salary in a company is $25,000, you can build a better picture of the salaries if you are also told that the range (spread) is $15,000–200,000. There are two main approaches to measuring spread.

Range. The range is intuitively appealing. The range of the series (2, 3, 5, 6, 24) is 2 to 24. Note how one extreme value (24) distorts the usefulness of the range. There are ways of lopping off such outliers. One is to discard the top and bottom few observations, perhaps the first and last 10% (deciles, see the box above).

An elaboration is the semi-interquartile range (sometimes called quartile deviation). This is one-half of the middle 50% of the distribu-

Calculating standard deviation 1

Standard deviation is simple if tedious to calculate without an electronic aid. For any set of numbers (say 2, 4, 6) the standard deviation (1.63 in this case) is found from the following six steps.

1 Find the arithmetic mean. $\qquad$ $(2 + 4 + 6) \div 3 = 4$
2 Find the deviations from the mean:
$$2 - 4 = -2$$
$$4 - 4 = 0$$
$$6 - 4 = +2$$
3 Square the deviations:
$$-2^2 = +4$$
$$0^2 = 0$$
$$+2^2 = +4$$
4 Add the results: $4 + 0 + 4 = 8$
5 Divide by the number of observations: $8 \div 3 = 2.67$
6 Take the square root of the result: $\sqrt{2.67} = \underline{1.63}$

Standard deviation is essentially the average of the deviations from the mean. Unfortunately, they will always add up to zero (eg, in the above example $-2 + 0 + 2 = 0$). Squaring the deviations disposes of negative signs since two negative numbers multiplied together always give a positive number. But the result is in squared units which is fairly silly. It is not helpful to refer to x dollars squared or y kilograms squared. So the final step is to take the square root of the average squared deviation.
Variance. The result at step 5 above (2.67) is known as the variance, a measure which is sometimes used in place of standard deviation in mathematical calculations.

tion. If the first quartile value is 14.5, and the third quartile is 27.5, the semi-interquartile range is $(27.5 - 14.5) \div 2 = 6.5$. This is not very satisfactory, because it uses only two observations. But it does cut off extreme values and it works well with skewed distributions.

Standard deviation. The standard deviation is a much better measure of spread. It uses every observation in a distribution and is not unduly influenced by one or two extreme values. It goes hand-in-hand with the mean, since they are both calculated on a similar basis. The standard deviation has not achieved the popularity of the mean because it is tedious to compute by hand (see Calculating standard deviation 1 box above), but it is easy to find with PC spreadsheet programs. Simply ask a PC spreadsheet for the standard deviation of the numbers in an identified spreadsheet range.

Interpreting standard deviation. As a very rough rule of thumb, two-thirds of a distribution is enclosed within one standard deviation on each side of the mean, 95% is captured within two standard deviations, and nearly everything is enclosed within three standard deviations in each direction. For example, if a machine produces metal rods, the length of which has a mean of 100mm and a standard deviation of 10mm, then 68% of rods are 90–110mm (100 ± 10) and 95% are 80–120mm (100 ± [10 × 2]) and 99.7% are 70–130mm (100 ± [10 × 3]).

Comparing standard deviations. When examining two distributions, it is sometimes useful to know how the spreads compare. For example, a manufacturing company might want to select the machine which offers the smallest variation in its output. If the mean outputs are the same, then obviously the company should pick the machine with the smallest standard deviation of output.

When comparing two standard deviations which are in different units, convert them both to percentages of the mean (known as coefficients of variation). For example, if French salaries have a mean of €50,000 and a standard deviation of €10,000, while comparable figures for Switzerland are SFr65,000 and SFr6,500, their coefficients of variation are 10,000 ÷ 50,000 × 100 = 20% and 6,500 ÷ 65,000 × 100 = 10%. France (coefficient of variation = 20%) has a much wider spread of salaries than Switzerland (10%).

Shape

The average shows where the mid-point of a distribution is located in the set of numbers (minus infinity to plus infinity). The standard deviation (or a similar measure) shows how widely a distribution is spread. Two distributions may have identical means and standard deviations, but totally different shapes. This is important, for example, to a butcher choosing between a pair of machines which both have a mean daily output of 5,000 sausages with standard deviation of 50 sausages. He might prefer to buy the one where output tends to be skewed above the mean rather than below; where, when there is variation, it will be in a favourable direction. There are two approaches to describing shape.

Skew and kurtosis. Skew (lack of symmetry) and kurtosis (peakedness) are two numerical measures which identify the shape of a distribution. They are not particularly useful in an everyday sense beyond being purely descriptive.

Standard distributions. A much more practical way of describing shape is with a standard distribution. Perhaps most useful and most common is the normal distribution. As will become clear, this is a powerful tool for analysing many business situations ranging from production and sales to financial risks. Conveniently, the techniques for dealing with the normal distribution apply to many other standard distributions (including the poisson, exponential and beta distributions discussed on pages 161–64 and 171). So if you know how the normal distribution works, you have access to a large number of analytical tools.

Normal distributions

If you know the mean and standard deviation of a set of data which conforms to a standard distribution, you have easy access to everything there is to know about the data. For example, Mr Ford has estimated that his used-car sales next year will average (ie, have a mean of) 10,000 vehicles with a standard deviation of 2,000 vehicles, and he knows that his sales follow a normal distribution. Based on no more than these three pieces of information, Mr Ford can derive figures such as the following:

- there is a 68% chance of selling 8,000–12,000 vehicles;
- there is a 95% likelihood of shifting 6,000–14,000 vehicles;
- there is a 7% probability of failing to sell the 7,000 vehicles required to break even.

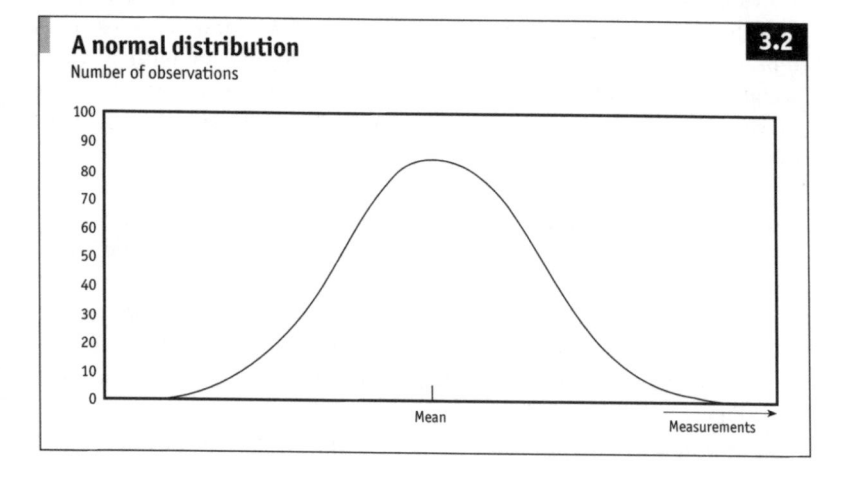

A normal distribution   **3.2**

Number of observations

How to recognise a normal distribution

Figure 3.2 is a graph of a normal distribution. What measurements are shown? Diameters of disks spinning off a production line? Heights of hardwood trees in Honduras? Yields per acre from mid-west wheat fields? The answer is all of these. Whenever a measurement is affected by a wide range of small independent factors, observations are distributed in this bell-shaped pattern.

The shape of the normal distribution is easy to understand if you think about anything which is affected by a large number of small independent influences, such as the output of a sausage machine. Most sausages will be very close to the average weight, although there are many which are just a touch lighter or heavier than average. The weights are arranged symmetrically around the mean weight. You are less likely to find a sausage of a given weight as you move further from the mean. The proportion of giant and midget sausages is very small. This symmetrical distribution with a declining frequency of observations further from the mean is the normal distribution.

The values written in the vertical and horizontal scales of the graph in Figure 3.2 will vary according to what is being measured. But if you draw most symmetrical distributions on a sheet of rubber and then stretch it in both directions you will be able to make the curve fit over the one in the illustration.

Analysing the normal

Rules of thumb. There are four facts which are easily derived for any normal distribution. As an example, consider Todd Teddies Inc's remarkable stuffing machine. It produces teddy bears with a mean weight of 2kg and a standard deviation of 0.2kg. The output of the machine is subject to a great many minor influences, so the weights of the bears can be taken as normally distributed.

- 50% of a normal distribution lies on either side of the mean, since the mean marks the mid-point. In this example, 50% of teddies weigh less than 2kg and 50% weigh more than 2kg.
- 68% of a normal distribution is contained between 1 standard deviation below the mean and 1 standard deviation above the mean. In this case, 68% of teddies weigh 1.8–2.2kg (ie, 2 ± 0.2). (See Figure 3.3.)
- 95% of a normal distribution lies between 2 standard deviations

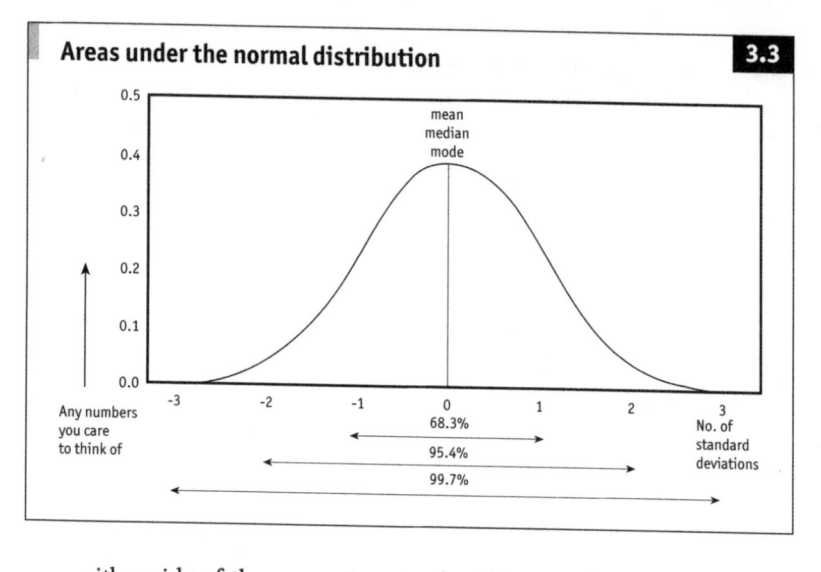

Areas under the normal distribution `3.3`

either side of the mean. So 95% of teddies weigh 1.6–2.4kg. (See Figure 3.3.)

- ▨ Nearly 100% of a normal distribution is enclosed within 3 standard deviations in each direction from the mean. Thus, nearly all of Todd teddies weigh 1.4–2.6kg. (See Figure 3.3.)

These rules of thumb help interpret standard deviations. Note the distinction between the actual standard deviation (0.2kg) and the number of these measured out from the mean (1, 2 or 3).

Normal tables. Other figures can be derived from the normal distribution. In fact, pick any point along the horizontal scale in Figure 3.3 and you can find out what proportions of observations are on either side of that point using tables of information relating to the normal distribution.

Another example will illustrate this. Mr Todd guarantees that his big bears will tip the scales at a minimum of 1.5kg each. How many teddies fall below this target weight? Recall that the mean weight is 2.0kg and the standard deviation is 0.2kg. The target is 0.5kg from the mean, or 2.5 times the standard deviation.

This allows the objective to be restated: any teddies weighing less than 2.5 standard deviations below the mean have to be slit ear to ear and restuffed.

Look at Table 3.2. Column A lists standard deviations. Run down these until you reach 2.5. Look across to the next column and read off

Table 3.2 **The normal distribution: z scores**

% distribution in the shaded areas

Z	In 1 tail	Outside 1 tail	In 2 tails	Between tails	Unit normal loss
A	B	C	D	E	F
0.0	50.00	50.00	100.00	0.00	0.399
0.1	46.02	53.98	92.04	7.96	0.351
0.2	42.07	57.93	84.14	15.86	0.307
0.3	38.21	61.79	76.42	23.58	0.267
0.4	34.46	65.54	68.92	31.08	0.230
0.5	30.85	69.15	61.70	38.30	0.198
0.6	27.43	72.57	54.86	45.14	0.169
0.7	24.20	75.80	48.40	51.60	0.143
0.8	21.19	78.81	42.38	57.62	0.120
0.9	18.41	81.59	36.82	63.18	0.100
1.0	15.87	84.13	31.74	68.26	0.083
1.1	13.57	86.43	27.14	72.86	0.069
1.2	11.51	88.49	23.02	76.98	0.056
1.3	9.68	90.32	19.36	80.64	0.046
1.4	8.08	91.92	16.16	83.84	0.037
1.5	6.68	93.32	13.36	86.64	0.029
1.6	5.48	94.52	10.96	89.04	0.023
1.7	4.46	95.54	8.92	91.08	0.018
1.8	3.59	96.41	7.18	92.82	0.014
1.9	2.87	97.13	5.74	94.26	0.011
2.0	2.28	97.72	4.56	95.44	0.008
2.1	1.79	98.21	3.58	96.42	0.006
2.2	1.39	98.61	2.78	97.22	0.005
2.3	1.07	98.93	2.14	97.86	0.004
2.4	0.82	99.18	1.64	98.36	0.003
2.5	0.62	99.38	1.24	98.76	0.002
2.6	0.47	99.53	0.94	99.06	0.001
2.7	0.35	99.65	0.70	99.30	0.001
2.8	0.26	99.74	0.52	99.48	0.001
2.9	0.19	99.81	0.38	99.62	0.001
3.0	0.14	99.86	0.27	99.73	0.000
3.1	0.10	99.90	0.20	99.80	0.000

Note: Where x is any value (see text), the percentage of a normal distribution on each side of x can be found against z above, where:

$$z = (x - \text{mean}) \div \text{standard deviation}$$
$$x = (z \times \text{standard deviation}) + \text{mean}$$
$$\text{standard deviation} = (x - \text{mean}) \div z$$

0.62. This indicates that 0.62% of a normal distribution lies beyond 2.5 standard deviations from the mean. Recall that 1.5kg is 2.5 bear standard deviations from the mean. So in other words, just 0.62% of teddies weigh less than 1.5kg; a very small proportion. Mr Todd has done his homework and pitched his guarantee at a level compatible with his equipment (and his local trading standards authority).

Interpreting Table 3.2. The table reveals what percentage of any normal distribution lies between two points. Read across from any standard deviation in Column A; say, 1.0 standard deviation (= 0.2kg in the bear example, where the mean was 2.0kg).

- ◪ Column B shows the proportion of the distribution beyond this distance from the mean; 15.9% of teddies weigh more than 2.2kg (2.0kg + 0.2kg). The picture at the top of the column illustrates this range. Since the normal distribution is symmetrical, the picture can be flipped left to right; there is 15.9% in each tail. So, also, 15.9% of teddies weigh less than 1.8kg (2.0kg – 0.2kg).
- ◪ Column C reveals the proportion of the distribution up to 1 standard deviation from the mean; 84.1% of teddies weigh less than 2.2kg (or, since the distribution is symmetrical, 84.1% weigh more than 1.8kg). It is not too hard to see that columns B plus C must equal 100%.
- ◪ Column D indicates the proportion outside a range on both sides of the mean, 31.7% of teddies weigh less than 1.8kg or more than 2.2kg. Since the hump is symmetrical, it is no surprise that the percentage in both tails (column D) is exactly twice the figure for one tail (column B).
- ◪ Column E is the result of some additional simple arithmetic. Since 31.7% of bears are outside of 1 standard deviation each way from the mean (column D), it follows that 68.3% are within this range (column E). That is, 68.3% of teddies weigh 1.8–2.2kg.
- ◪ Column F is slightly different. It relates to a special use of the normal distribution for estimating potential profit and loss figures. It is included here so that one table contains all the information relating to the normal but it is not discussed until pages 153ff.

z scores. By convention, the number of standard deviations listed in column A are known as z scores. For example, at 2.5 standard deviations, $z = 2.5$. Any measurement (call it x) can be converted into a z score by a simple relationship: $z = (x - mean) \div$ standard deviation. In the example above, where the mean weight of the teddies was 2.0kg, the standard deviation was 0.2kg and Mr Todd was interested in the cut-off point $x = 1.5$kg, the z score relationship reveals that $z = (1.5 - 2.0) \div 0.2 = -0.5 \div 0.2 = -2.5$. (The minus sign indicates that the target z is 2.5 standard deviations below the mean. A positive z indicates a standard deviation above the mean.)

Starting from a percentage. Of course it is not compulsory to enter the table from the first column. It is often useful to start with a percentage. Suppose Mr Todd decides to enhance his marketing by introducing a new range of giant bears. He decrees that the heaviest 20% of his teddies will take a new label. What is the cut-off weight? He is interested in one tail of the distribution. Run down column B to locate a figure close to 20%. The nearest is 21.2. This is next to $z = 0.8$, which has to be converted back into kilograms. The relationship used above can be rearranged into $x = (z \times$ standard deviation$) +$ mean $= (0.8 \times 0.2) + 2.0 = 2.16$. Thus, the heaviest 20% of bears weigh about 2.16kg or more.

Filling in the gaps. Table 3.2 is not detailed enough to be exact. In the example just given, the result was about 2.16kg. You can improve on this by consulting a more detailed set of statistical tables or applying logic. The logic is simple and often accurate enough. In this example, the one-tail proportions nearest to 20% are 21.2% at 0.8 standard deviations and 18.4% at 0.9 standard deviations. As 20% is not quite halfway between the two, call the required standard deviation 0.84. The difference is not

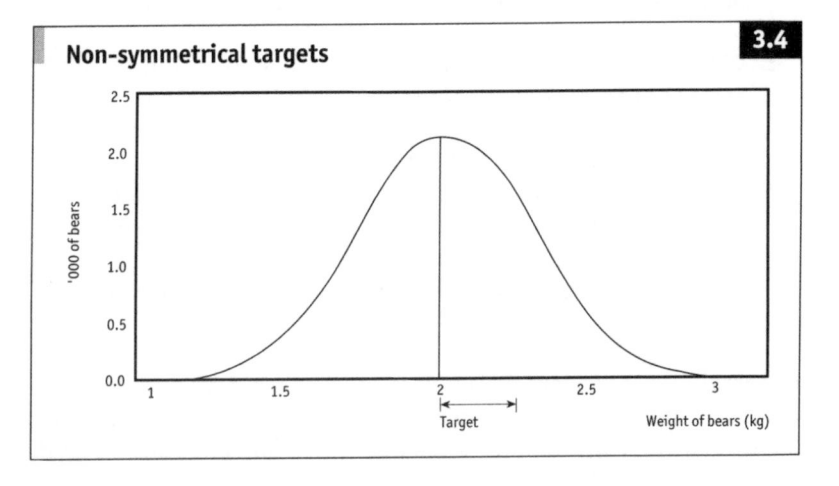

Non-symmetrical targets 3.4

great: $(0.84 \times 0.2) + 2.0 = 2.17$kg compared with the 2.16kg previously obtained. Such estimation is called interpolation.

Non-symmetrical targets. So far Table 3.2 has been used to derive information about the distribution in relation to the mid-point (the mean). One final example will illustrate a still wider application.

Mr Todd decides to examine all bears weighing between 2.1 and 2.4kg. What proportion of the total is this? Figure 3.4 illustrates the target. There is not a matching column in Table 3.2. But column C covers the area to the left of the target, and column B covers the area to the right. Add these two together to find out how much of the distribution is outside of the target range. Subtract that from 100% to reveal how much is within target.

First, use the established relationship to convert the weights into z scores: $z_1 = (2.1 - 2.0) \div 0.2 = 0.5$ and $z_2 = (2.4 - 2.0) \div 0.2 = 2.0$.

Column C reveals that 69.15% of teddies weigh less than 0.5 standard deviations from the mean. Column B indicates that 2.28% weigh more than 2.0 standard deviations from the mean. So $69.15 + 2.28 = 71.43\%$ of bears are outside of the target range, and $100 - 71.4 = 28.6\%$ are within the target. For every 1,000 bears produced by Mr Todd's sweat shop, he will examine 286.

All this information about bears is derived from just two figures (mean and standard deviation) and the knowledge – or assumption – that the teddies' fillings are distributed normally.

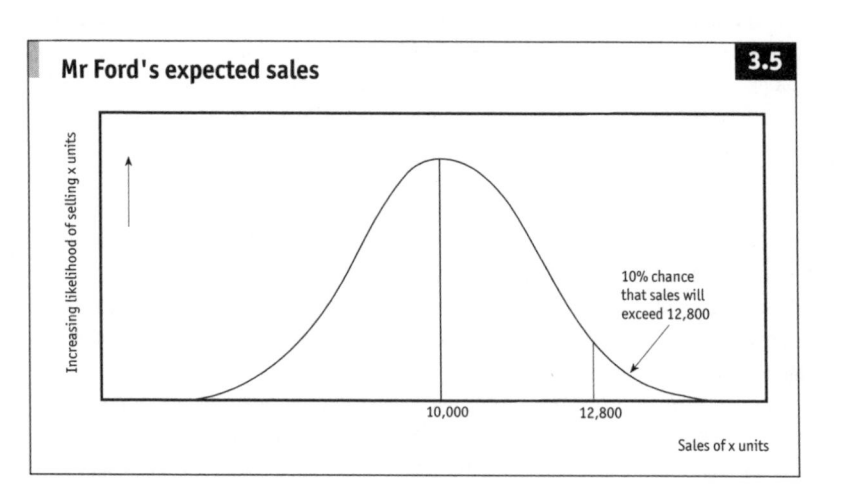

Mr Ford's expected sales 3.5

Increasing likelihood of selling x units

10% chance that sales will exceed 12,800

10,000   12,800

Sales of x units

Modelling risk with the normal distribution

The normal distribution has countless uses beyond the analysis of observed data. It is extremely good for modelling risk. Think of the normal curve (eg, Figure 3.5) as tracing the distribution of possible outcomes for some event such as the level of next year's sales. The most likely outcome (the mean) is the highest point on the graph. The standard deviation measures how widely the risks are spread. As you move further away from the mean, the probability of any outcome reduces.

For example, sales can often be assumed to be normally distributed. Mr Ford has estimated that his used-car sales next year will average 10,000 with a standard deviation of 2,000 vehicles (more on how he obtained these figures in a moment). Since 68% of a normal distribution is within 1 standard deviation on either side of the mean, Mr Ford knows that there is a 68% chance that he will sell between 8,000 and 12,000 vehicles (10,000 ± 2,000). Suppose that he needs to shift 7,000 to break even. The proportion of the probability distribution below (ie, to the left of) 7,000 vehicles indicates the chances of failing to sell that many.

The basic relationship, $z = (x - mean) \div$ standard deviation, gives $(7,000 - 10,000) \div 2,000 = -1.5$. From Table 3.2, where $z = 1.5$, the amount in the left-hand tail of the distribution is 6.68%. Based on these figures, Mr Ford has a mere 7% risk of failing to break even.

Quantifying risk. The mean level of sales is simply the most likely sales figure. If Mr Ford states "I expect to sell 10,000 cars next year" he is implying that his forecast mean is 10,000. One other prediction will produce a standard deviation. For example, Mr Ford estimates that he has no more than a 10% chance of selling in excess of 12,800 vehicles. The basic z score relationship indicates that: standard deviation = $(x - mean) \div z$. For Mr Ford's figures, where the proportion is 10%, $z = 1.28$. So the standard deviation of his expected sales is $(12,800 - 10,000) \div 1.28 = 2,000$.

This presupposes that the forecasting is accurate, and that probable sales follow a normal symmetrical pattern. Nevertheless, such analysis provides a useful input to the decision-making process.

Financial risk. If you assess the risks of any financial investment to be symmetrical, the arithmetic relating to the normal distribution can be used for investment analysis.

For example, a punter makes the heroic assumption that stock market prices are just as likely to rise as they are to fall. Say this person estimates that in two months a share price is likely to be $100 but there is a

15% chance that it will be $150. From these, the standard deviation of the risk can be found with a simple rule of thumb: 15% of a normal distribution is beyond one standard deviation from the mean. So one standard deviation is $50, which is the 15% figure ($150) less the mean ($100).

Once more, from the mean ($100) and standard deviation ($50), other figures follow. The punter could, for example, say that there is a 40% chance that the price will fall below $87.50: from Table 3.2, for 40% in the left-hand tail, $z = -0.25$. Using $x = (z \times \text{standard deviation}) + \text{mean}$ gives $(-0.25 \times 50) + 100 = \$87.50$.

How such forecasts are made and how the derived probabilities are incorporated into decision-making are discussed in following chapters.

4 Tables and charts

> "A statistician is a person who deals with numbers and charts,
> but who doesn't have the personality to become an
> accountant."

Anon

Summary

While measures such as averages and totals (see Chapter 3) summarise data within one or two figures, tables and charts bring order while still presenting all the original information. Tables and charts are powerful aids to interpretation. Tables are generally capable of conveying information with a greater degree of precision than charts, but charts can capture far more information in one go. Charts are exceptionally good at getting a message across, especially in weighty reports and presentations.

Just a few numbers can be expanded into a huge table to aid analysis and review (see Table 4.1, page 76). Spreadsheets make this especially easy, but shrink the tables ruthlessly before putting them on general release; otherwise the mass of numbers will confuse the uninitiated. Summary measures, such as row or column totals and averages, help build the overall picture and actual or percentage changes are often more helpful than the raw data.

When faced with graphs, always check the vertical scale to make sure that it has not been expanded to make a small change look large or, worse, inverted to make a fall look like a rise. Then think about what might have been carefully omitted. Two points showing healthy year-end financial balances might be joined by a straight line which ignores a plunge into the red in the intervening period.

As with tables, charts of actual or percentage changes over 12 months are helpful for analysing monthly figures which have a distracting seasonal pattern (see Chapter 5). With index numbers, recall that the convergence illusion can be misleading (see Figure 1.2, page 15).

Tables

There are two main types of table: one used when presenting data, and one for interpretation.

Presenting data

Tables for use by others should follow the rule of three. There are three objectives, each with three tricks.

1 Make tables concise by eliminating non-essential information (be ruthless); rounding to two or three significant figures (including, preferably, no more than one decimal place) and converting to index form if the original numbers are clumsy; and omitting grid lines and avoiding too much white space (which makes it hard to compare related data).

2 Make tables informative by including summary measures (perhaps row and column totals or averages, and/or percentage changes or breakdowns); incorporating clear and complete labels; and adding a two-line written summary.

3 Order tables in columns (it is easier to follow a sequence of numbers down a column, rather than across a row); by importance (do not bury the important columns/rows in the middle of the table); and by size (arrange figures according to orders of magnitude in the key rows/columns).

Note how the weekly tables at the back of *The Economist* follow these rules (within the constraints of house style). The presentations often exclude raw data altogether in favour of percentage changes. For example, instead of publishing the crude index of industrial production in the 15 developed countries and the euro area which it follows each week, *The Economist*'s tables show the percentage change over the past 12 months making it easy to spot the fastest or slowest growing economy over the past year.

Interpreting data

Preparing numerical information for analysis is not unlike preparing a tabular presentation. However, in this case the inclusion of additional information is required. Spreadsheets are wonderful; with little effort by the user they add percentage breakdowns, percentage changes, absolute changes, totals, averages, and so on. These are the numerical route maps that lead to the important patterns in data.

Table 4.1 shows this tabular analysis in action. The 16 figures enclosed in the small box (top left) were provided. All the other numbers were derived from these 16. There is no need to describe everything that the table reveals, but note how the index numbers are easier to follow, how the index of imports relative to (divided by) GDP shows

Table 4.1 **Tabular analysis**

	GDP $m	Imports $m	Imports relative to GDP	GDP index	Imports index	Imports as % of GDP
2001						
1Qtr	98,249	25,784	100.0	100.0	100.0	26.2
2Qtr	99,251	27,645	106.1	101.0	107.2	27.9
3Qtr	106,364	29,662	106.3	108.3	115.0	27.9
4Qtr	112,973	29,276	98.7	115.0	113.5	25.9
2002						
1Qtr	108,962	28,718	100.4	110.9	111.4	26.4
2Qtr	109,874	30,875	107.1	111.8	119.7	28.1
3Qtr	118,278	32,959	106.2	120.4	127.8	27.9
4Qtr	124,763	32,642	99.7	127.0	126.6	26.2
TOTAL						
2001	416,837	112,367	102.7	106.1	109.0	27.0
2002	461,877	125,194	103.3	117.5	121.4	27.1

Changes					GDP	Imports
	%	%	%		$m	$m
Over 1 quarter						
2001						
2Qtr	1.0	7.2	6.1		1,002	1,861
3Qtr	7.2	7.3	0.1		7,113	2,017
4Qtr	6.2	-1.3	-7.1		6,609	-386
2002						
1Qtr	-3.6	-1.9	1.7		-4,011	-558
2Qtr	0.8	7.5	6.6		912	2,157
3Qtr	7.6	6.7	-0.8		8,404	2,084
4Qtr	5.5	-1.0	-6.1		6,485	-317
Over 4 quarters						
2001–02						
1Qtr	10.9	11.4	0.4		10,713	2,934
2Qtr	10.7	11.7	0.9		10,623	3,230
3Qtr	11.2	11.1	-0.1		11,914	3,297
4Qtr	10.4	11.5	1.0		11,790	3,366
Over 1 year						
2001–02	10.8	11.4	0.6		45,040	12,827

how one series is moving in relation to the other, and how annual figures indicate so much about overall trends. Lastly, note that the raw figures are not seasonally adjusted, so the changes over four quarters are especially valuable.

The first step is always to decide what you want to find out from the data. Then look for the patterns; they will reveal the answers.

Charts

A chart is an immensely powerful way of presenting numerical data. All the information is summarised in one go, in a way that the eye can readily absorb. Trends, proportions and other relationships are revealed at a glance.

Pictorial accuracy

There is some inevitable loss of accuracy with charts, depending on size, scale and the number of points plotted. This seldom outweighs the impact of visual presentation. Where underlying values are to be communicated, they can be written on the graph if they are few, or listed in an accompanying table. Remember, though, that too much information on one chart can be confusing.

Scales

Range. Always check the scales on a chart before you look at the central area. Graphs are usually most revealing when the vertical axis covers only the range of values plotted, rather than starting at zero, although the relative movements are altered.

Inverted vertical scales. Charts that increase in value towards the top are familiar. Inverting the y-scale is the easy way to make declining

---

Anatomy of a graph

By convention, the independent variable – often time – is known as x and drawn on the horizontal x-axis of a graph. The variable that depends on x, the dependent variable y, runs up the vertical y-axis. Points on the graph are identified by their measurements along the x- and y-axes. This is written (x, y) for short. The point (2,1) below indicates that x = 2 and y = 1. These are called Cartesian co-ordinates after the French philosopher René Descartes. A line joining a sequence of points is known as a curve, even if it is straight.

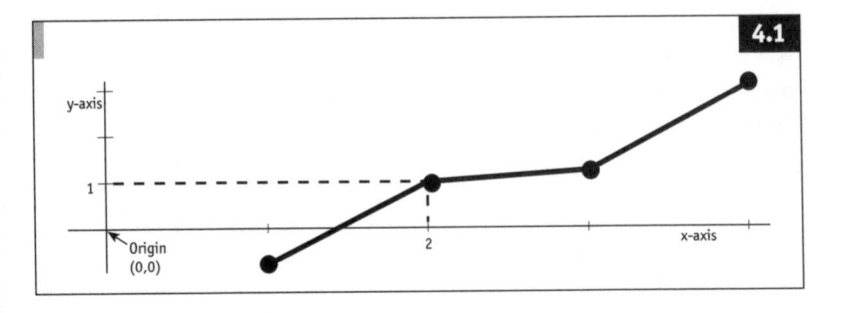

profits look presentable (a better way to present worse results). This trick is more legitimate when trying to avoid the impression that a currency, say, is falling when it is not (Figure 4.3) or when comparing series that move in opposite directions.

Joining the dots
Points plotted on a graph may be joined together with ragged, stepped, or smooth lines (known, misleadingly, as curves). Purists argue that only continuous data should be linked in this way (see Figure 4.4). A straight line between two points may conceal an uncharted journey. Usually, though, it is meaningful to link the points in some way. This may be justifiable as it reveals trends at a stroke.

Vertical range
There is only about 3% separating the highest and lowest values on this chart, even though there appears to be wide variation between them at first glance. Doubling the height of a bar represents a doubling in the magnitude it represents only if the vertical scale starts at zero. If the vertical scale did start at zero, the tops of the bars would be almost level and the chart would be fairly meaningless. Breaking the scale draws attention to the fact that the scale does not start at zero.

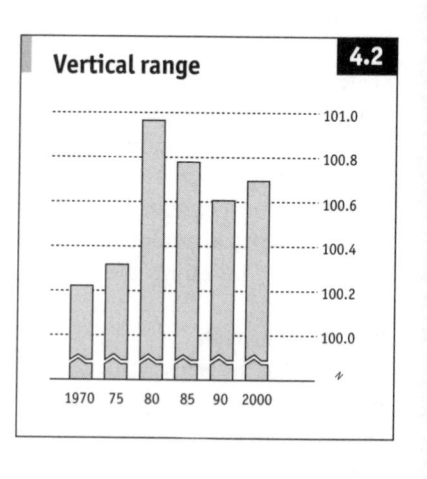

Inverted currencies
Currencies are often quoted in terms of the US dollar (ie, so many Australian dollars or euros equal 1 US dollar). A decline in the numerical value of the exchange rate to the dollar represents a rise in the value of a currency against the dollar. The best way to chart the euro's progress is to invert the scale on the graph, as here. The scale would be shown the other (normal) way up if the aim was to show the rise and fall of the US dollar.

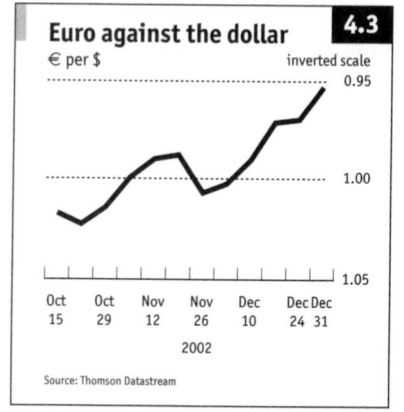

Scatters, lines and bars

Figures 4.5–4.15 show the key types of chart: scatter plots, line graphs, bar charts and pies.

A misleading line of enquiry

Graphs may not tell the whole truth. The unbroken line in this figure runs from A to B, implying that intermediate values can be read from the graph. The dotted line reveals the actual path of the data. The divergence might be even more serious where, for example, a graph shows a stream of healthy end-year financial balances and omits to show that the balances plunged into deficit in the intervening months.

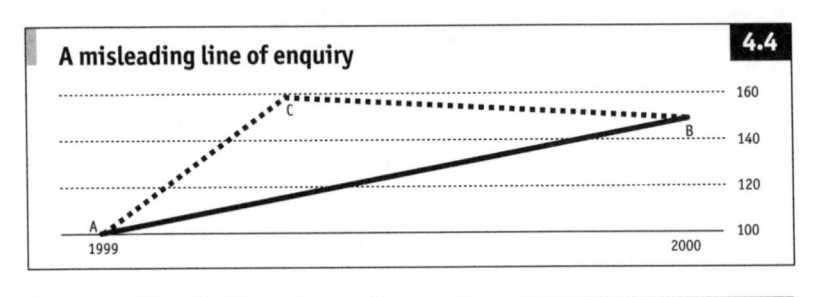

4.4

A misleading line of enquiry

A scatter graph that makes a point

This sort of chart, called a scatter diagram for obvious reasons. Its strength is that it clearly reveals relationships between the points. It is frequently helpful in the early stages of data analysis.

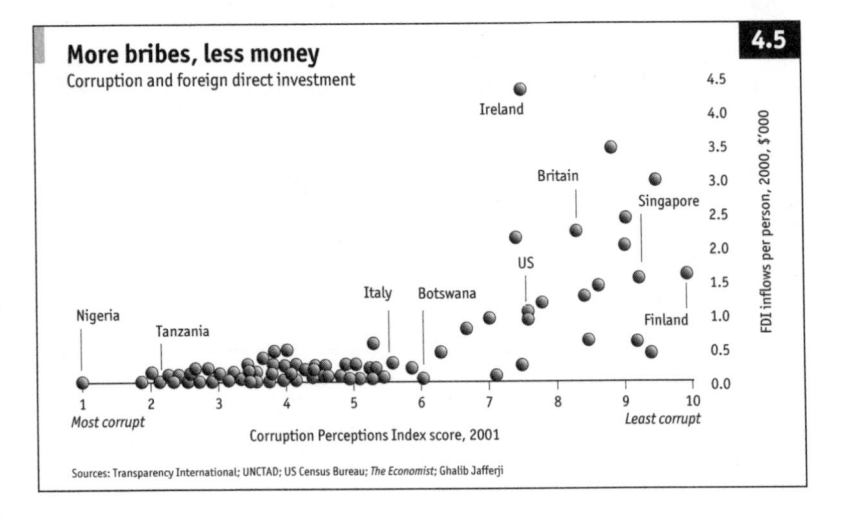

4.5

More bribes, less money

Corruption and foreign direct investment

Sources: Transparency International; UNCTAD; US Census Bureau; *The Economist*; Ghalib Jafferji

A relational diagram

This chart shows the relationship between wage increases and the unemployment rate, and gives a good indication of the business cycle.

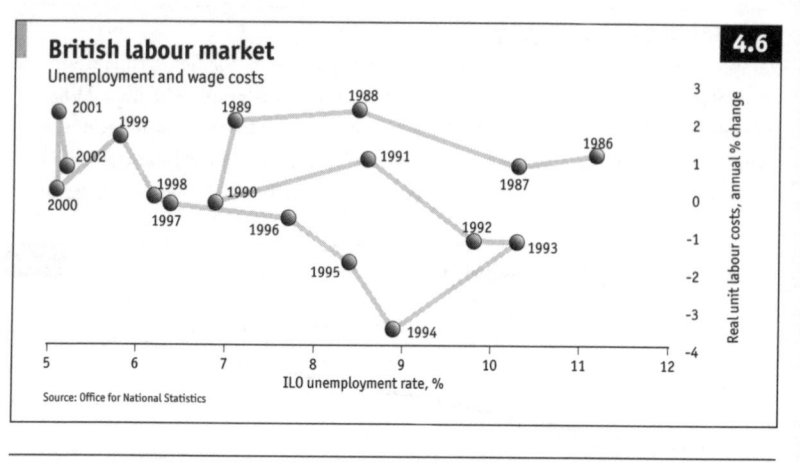

British labour market 4.6
Unemployment and wage costs
Real unit labour costs, annual % change
ILO unemployment rate, %
Source: Office for National Statistics

A simple line graph

Simple line charts, where points are joined with lines, allow tens or even hundreds of values to be presented at once. The lines help organise the relationship between the points and reveal trends at a stroke. Note how stepped lines are sometimes appropriate. Figure 4.7b illustrates a way of combining on one chart series of numbers with different units by converting all the series to indices with a chosen numerical value equal to 100. But remember how indices misleadingly appear to converge on the base.

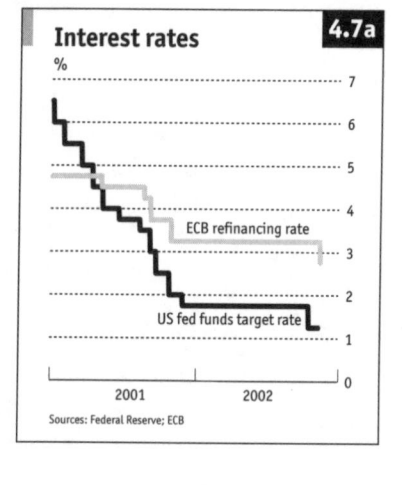

Interest rates 4.7a
%
ECB refinancing rate
US fed funds target rate
2001   2002
Sources: Federal Reserve; ECB

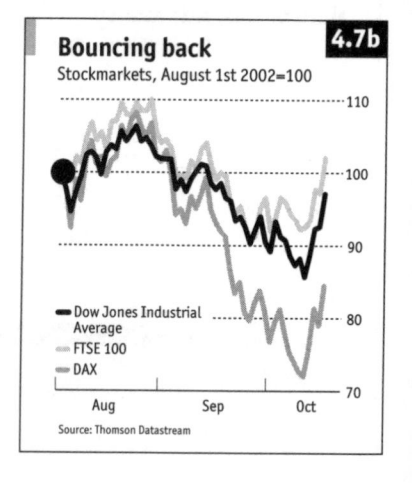

Bouncing back 4.7b
Stockmarkets, August 1st 2002=100
Dow Jones Industrial Average
FTSE 100
DAX
Aug   Sep   Oct
Source: Thomson Datastream

High-low

The lines represent a range of in this instance commodity prices. The top and bottom ticks represent highs and lows for the period while the bars indicate change to the closing value. The high-low chart is a favourite of Wall Street analysts. Stock prices are often shown as vertical lines on a time scale – the range for the day – with a little horizontal tick to show the closing value.

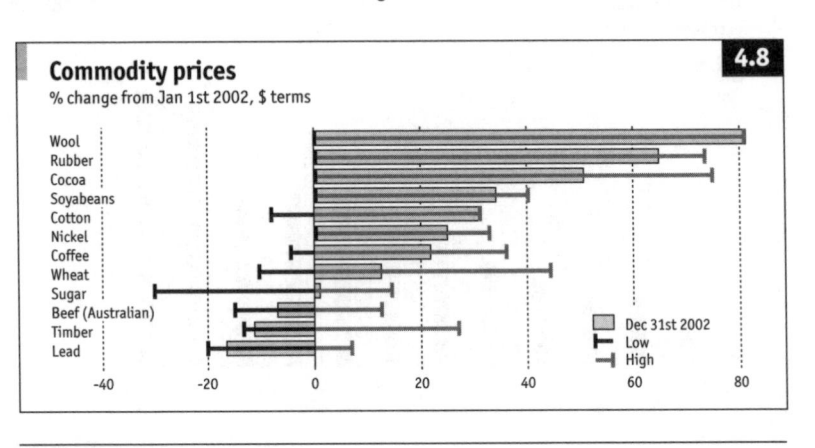

Commodity prices
% change from Jan 1st 2002, $ terms

4.8

Wool, Rubber, Cocoa, Soyabeans, Cotton, Nickel, Coffee, Wheat, Sugar, Beef (Australian), Timber, Lead

Dec 31st 2002
Low
High

-40 -20 0 20 40 60 80

Line up

Conventional joining-the-dots is not appropriate when points are not directly interrelated. Here, vertical lines distinguish between points and highlight those which are linked. Relative magnitudes and changes are easy to spot. This is almost a bar chart.

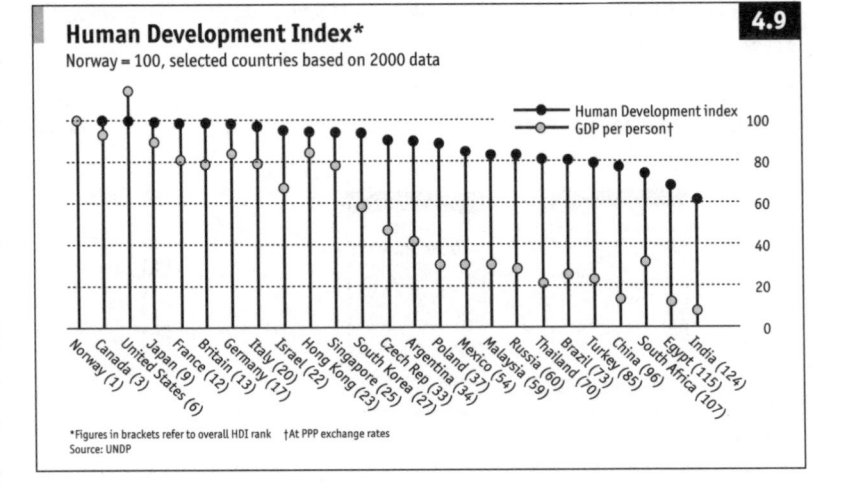

Human Development Index*
Norway = 100, selected countries based on 2000 data

4.9

Human Development index
GDP per person†

Norway (1), Canada (3), United States (6), Japan (9), France (12), Britain (13), Germany (17), Italy (20), Israel (22), Hong Kong (23), Singapore (25), South Korea (27), Czech Rep (33), Argentina (34), Poland (37), Mexico (54), Malaysia (59), Russia (60), Thailand (70), Brazil (73), Turkey (85), China (96), South Africa (107), Egypt (115), India (124)

*Figures in brackets refer to overall HDI rank   †At PPP exchange rates
Source: UNDP

Grouped bars

Bar charts are appropriate where data are obviously discrete or where they represent categories. Grouping spotlights changes within each category. Sometimes the bars are represented horizontally rather than vertically.

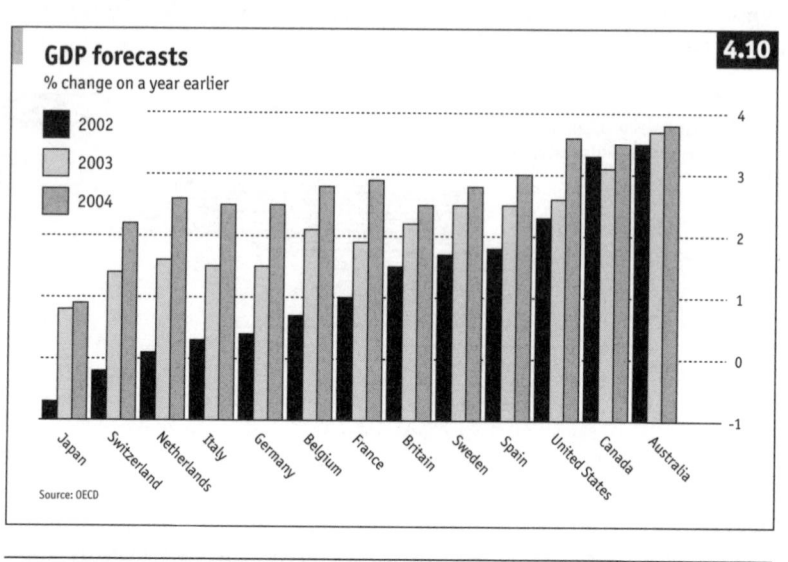

GDP forecasts
% change on a year earlier

4.10

■ 2002
☐ 2003
▨ 2004

Japan Switzerland Netherlands Italy Germany Belgium France Britain Sweden Spain United States Canada Australia

Source: OECD

Stacked bars

Stacked bars reveal total magnitudes and break-downs. But it can be difficult to compare components from bar to bar (except for the one starting from the horizontal axis). Stacked bars are a good place to hide the odd poor result among generally good figures.

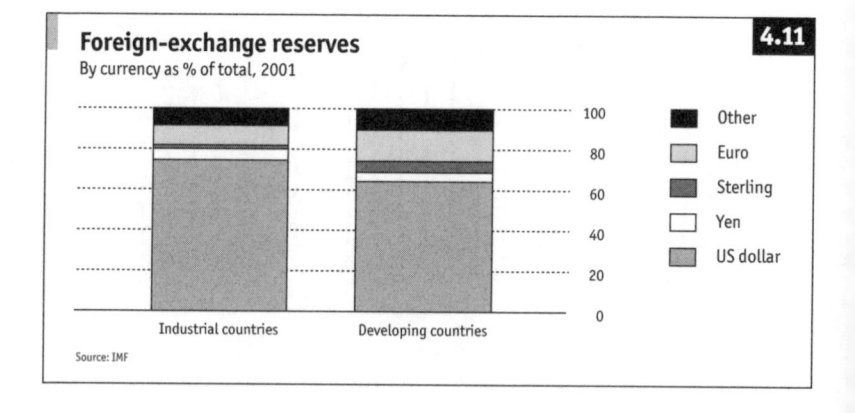

Foreign-exchange reserves
By currency as % of total, 2001

4.11

■ Other
☐ Euro
▨ Sterling
☐ Yen
▨ US dollar

Industrial countries   Developing countries

Source: IMF

Relatives

A useful trick when comparing two items, such as two share prices (left) is to straighten one out and see how the other moves in relation or relative to that (right). For each period, simply divide the value for the second series (Stuckfast) by the one for the series to be straightened (Superstick).

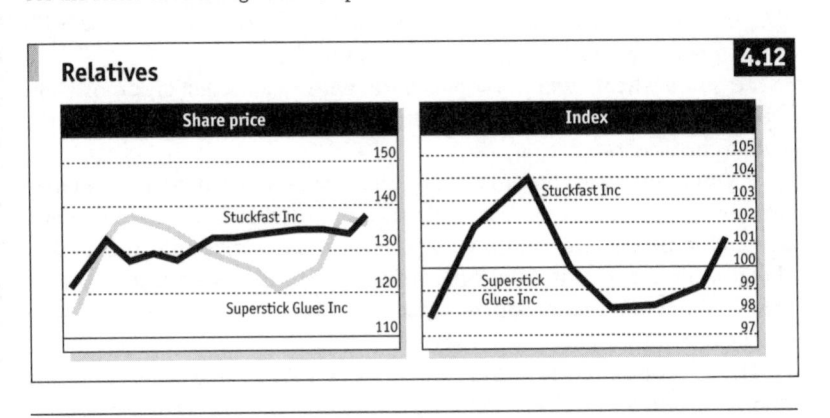

Relatives 4.12

Continuous area

This graph is a hybrid; a cross between a line chart and a stacked bar chart. It shows the world ecological footprint and the breakdown by type over a longer period than would be feasible with a bar chart. It highlights trends and breakdowns at the same time.

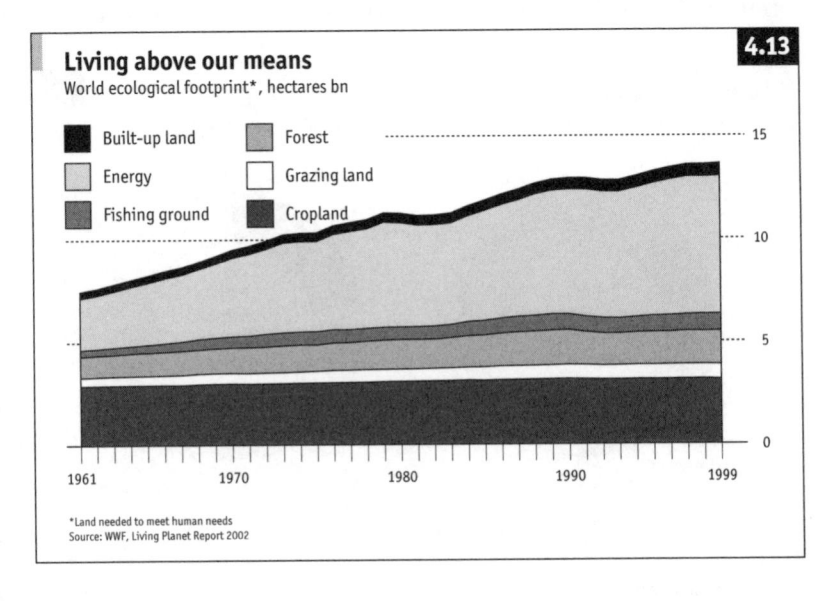

Living above our means 4.13
World ecological footprint*, hectares bn

*Land needed to meet human needs
Source: WWF, Living Planet Report 2002

Pie charts

Pie charts are an alternative to stacked bars for illustrating breakdowns in cross-sectional data.

Each slice of the pie indicates the proportion of the whole taken up by the item allocated to that segment. For example, the pies in Figure 4.14 indicate that America and Russia each produce 10% of the world's oil and respectively 23% and 22% of its natural gas.

PC spreadsheets will draw pie charts with ease. Select graph or chart from the menu, indicate that a pie chart is required, and identify the data to be illustrated.

If you draw a pie chart manually, share out the 360 degrees in a circle among the components. For each segment, divide by the total represented by the pie and multiply by 360 degrees. For example, the oil production pie represents total production of 74.5m barrels a day, out of which Saudia Arabia produces 8.94m barrels. So Saudi Arabia is 8.94m ÷ 74.5m = 0.12 × 360 = 43.2 degrees.

The area of a circle is found from $\pi \times r^2$, where r is the radius (half the diameter) and π (the Greek letter pi) is the constant 3.14159. For example, a circle with a radius of 5cm has an area of $3.14159 \times 5^2 = 3.14159 \times 5 \times 5 = 78.5cm^2$.

Too many segments will overcrowd pie charts. But they take well to annotation. An approach using a combination of pies and tables is often useful as it shows percentages as well as actual data, and

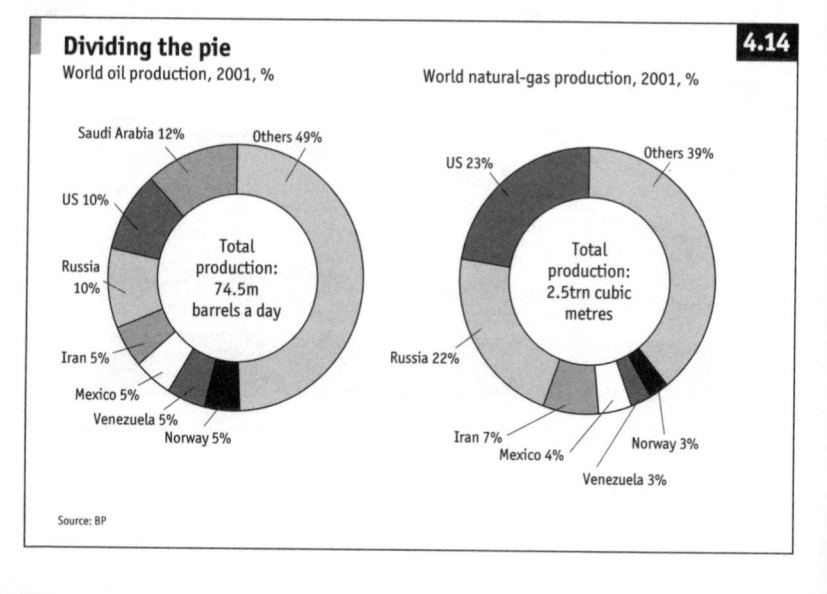

Dividing the pie `4.14`

World oil production, 2001, %

World natural-gas production, 2001, %

Saudi Arabia 12%   Others 49%

US 10%

Russia 10%

Total production: 74.5m barrels a day

Iran 5%

Mexico 5%

Venezuela 5%

Norway 5%

US 23%   Others 39%

Russia 22%

Total production: 2.5trn cubic metres

Iran 7%

Mexico 4%

Venezuela 3%

Norway 3%

Source: BP

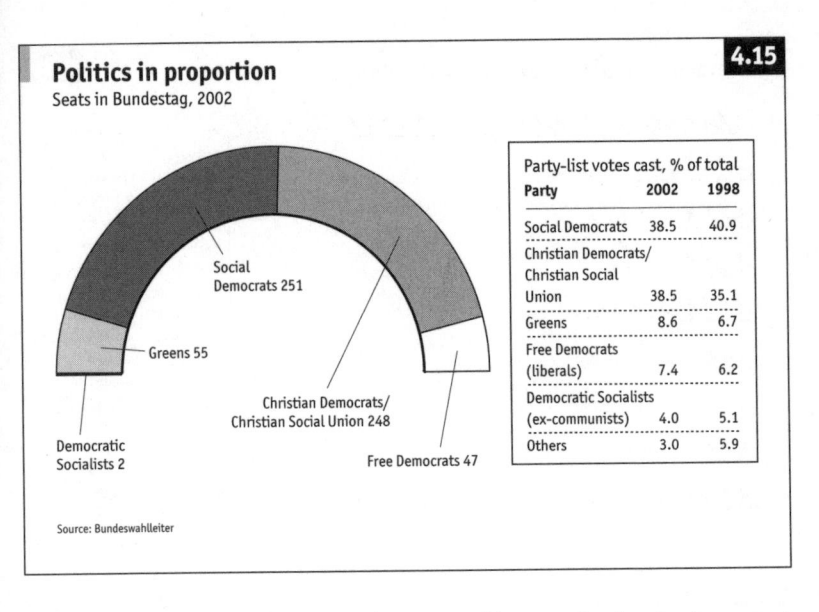

Politics in proportion
Seats in Bundestag, 2002

4.15

Social Democrats 251

Greens 55

Christian Democrats/
Christian Social Union 248

Democratic
Socialists 2

Free Democrats 47

Party-list votes cast, % of total		
Party	**2002**	**1998**
Social Democrats	38.5	40.9
Christian Democrats/ Christian Social Union	38.5	35.1
Greens	8.6	6.7
Free Democrats (liberals)	7.4	6.2
Democratic Socialists (ex-communists)	4.0	5.1
Others	3.0	5.9

Source: Bundeswahlleiter

draws attention to a few important numbers at the level of greatest detail. For example, Figure 4.14 could include tables giving the number of barrels of oil and cubic metres of natural gas produced by different countries, including some of those accounted for in the "others" category.

Variations on a pictorial theme
Where information does not lend itself to a graphical comparison of magnitudes, other pictorial presentation is often effective. Figures 4.15 to 4.17 show data that would be less interesting in tabular form. Such presentations are limited only by the imagination of the person producing or specifying the format. Clever computer software is opening up such techniques to non-artists.

A pictogram
This pictogram is essentially a bar chart. Relative magnitudes are represented by the lengths of the sequences of pictures. But many pictograms, such as two barrels showing oil consumption in two countries, are pies in disguise where areas are indicative of relative size. Including tabular information as well gives a much fuller picture.

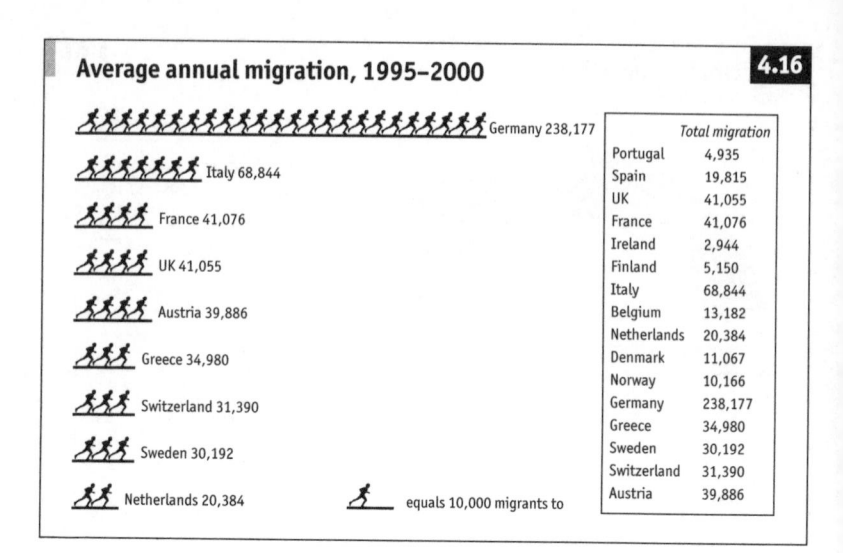

Average annual migration, 1995–2000   4.16

Germany 238,177

Italy 68,844

France 41,076

UK 41,055

Austria 39,886

Greece 34,980

Switzerland 31,390

Sweden 30,192

Netherlands 20,384

equals 10,000 migrants to

	Total migration
Portugal	4,935
Spain	19,815
UK	41,055
France	41,076
Ireland	2,944
Finland	5,150
Italy	68,844
Belgium	13,182
Netherlands	20,384
Denmark	11,067
Norway	10,166
Germany	238,177
Greece	34,980
Sweden	30,192
Switzerland	31,390
Austria	39,886

A diamond chart

Another possibility is the diamond, based on four different scales, used effectively here to compare three countries against the Maastricht criteria.

Out of the frame   4.17
Maastricht criteria, %

Czech Republic

Maastricht

Hungary

Poland

10 *Inflation rate*

10
Interest rate

10
*Deficit, %
of GDP*

100 *Debt, % of GDP*

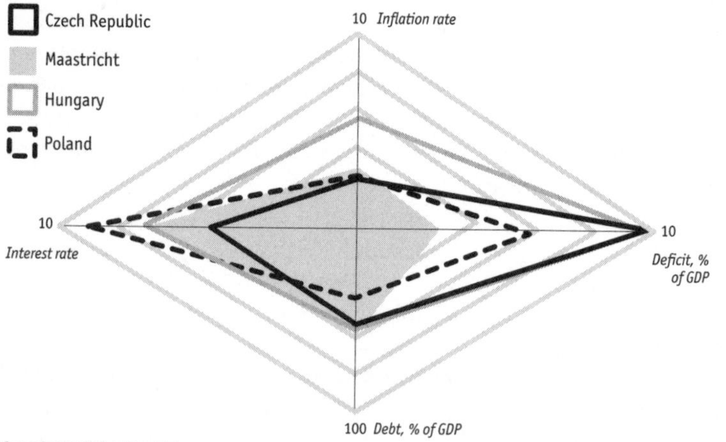

Source: Dresdner Kleinwort Wasserstein

How to mislead

Introduce a statistician to a cartographer and there is no limit to their visual wizardry. Never mind if the data do not fit the argument, a chart will soon prove the point – eg, scales that do not start from zero will exaggerate modest movements, while logarithmic scales can flatten out an increase. A logarithmic scale is legitimate if comparing rates of growth, but if it is the absolute level that matters, stick to an ordinary scale.

Suppose the national union of office cleaners was after a big pay rise, then in Figure 4.18 the top left chart (on an ordinary scale) would make impressive reading: over the past ten years, the salary of managers appears to have risen much faster than that of office cleaners, and the wage gap has widened. Managers might offer the top right chart (on a logarithmic scale) in their defence. This suggests (correctly) that wages of office cleaners have been rising at a faster pace than managers', and (misleadingly) that the absolute wage gap has narrowed.

Even so, the absolute gap still looks embarrassingly large. No problem; rebase the figures on to an index (bottom left chart) or express them as a percentage change on a year ago (bottom right chart) – ie, pick the chart that suits you best.

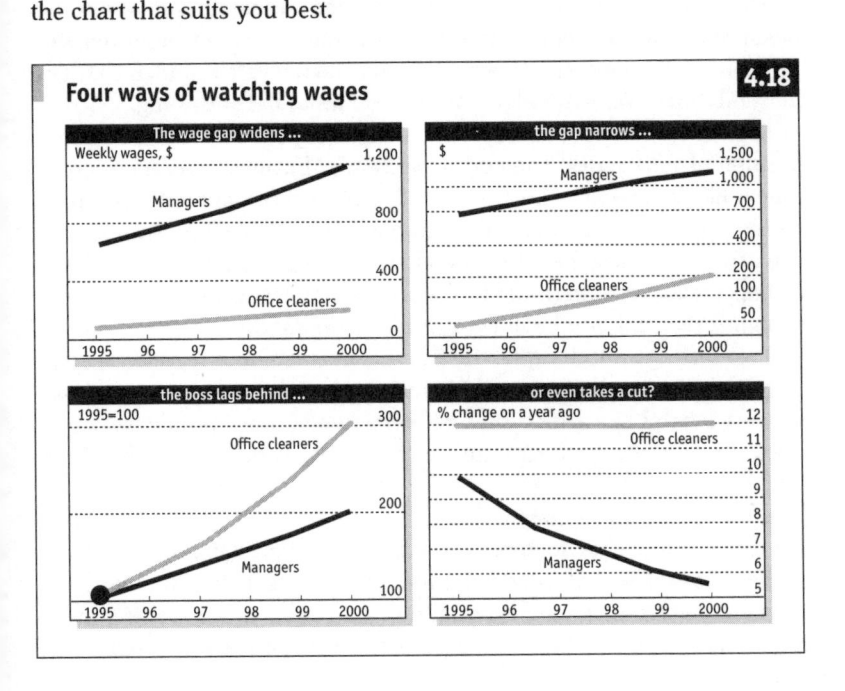

Four ways of watching wages **4.18**

5 Forecasting techniques

"Forecasting is very dangerous, especially about the future."

Samuel Goldwyn

Summary

Forecasting is fundamental to business decision-making. There are three main methods:

- Subjective forecasting is based on experience, intuition, guesswork and a good supply of envelope-backs.
- Extrapolation is forecasting with a ruler where past trends are simply projected into the future.
- Causal modelling (cause and effect) uses established relationships to predict, for example, sales on the basis of advertising or prices.

The main message is that nothing is certain. Whether you are in the business of extrapolation or causal forecasting, your predictions depend on more of the same. Numerical methods will not catch sudden changes in trends. Management judgment is a vital input to forecasting.

For extrapolation or causal forecasting, a key idea concerns the composition of time series. The performance of a business over a period of time is likely to reflect a long-run trend, a cyclical component, a seasonal pattern and a harder to explain random residual element. No business can afford to ignore these factors when thinking about the future. Identifying a seasonal pattern can be especially important, and it is relatively straightforward if you have enough historical data.

Two important techniques are moving averages and regression analysis. Both are invaluable tools for forecasting, and regression analysis, because it helps to identify links between two or more variables such as sales and advertising expenditure, makes decision-making much tighter.

The importance of judgment

All three approaches to forecasting are important. Quantitative techniques – causal modelling and extrapolation – are exciting. But they tend to acquire false authority, especially if the results are presented in an impressive computer printout. Forecasts are not necessarily right, just because they are based on numerical techniques. Moreover, a forecasting method which worked last year is not automatically reliable this time round. Quantitative forecasts rarely take account of every relationship and influence and even the most comprehensive and rigorous projection can be ruined by capricious human behaviour. Consequently, forecasts should always be tempered with a good dose of managerial judgment, even if only to say that a forecast is acceptable.

Time series

Forecasting, whether by extrapolation or causal modelling, is about predicting the future. It must involve time series. Even if a forecast is apparently cross-sectional, such as sales by all branches in one period, it is a cross-section of a time series. It is a good idea to start by looking closely at time series.

Types of time series?

Units. Time series come in all sorts of units: the dollar value of ice cream sales each day, the number of cars produced each month, the tonnage of coal produced every year. Also, all units can be converted into index form (see page 14).

Volume, value and price. For those who like to put things in neat little boxes, any time series can be pigeon-holed into one of three classifications,

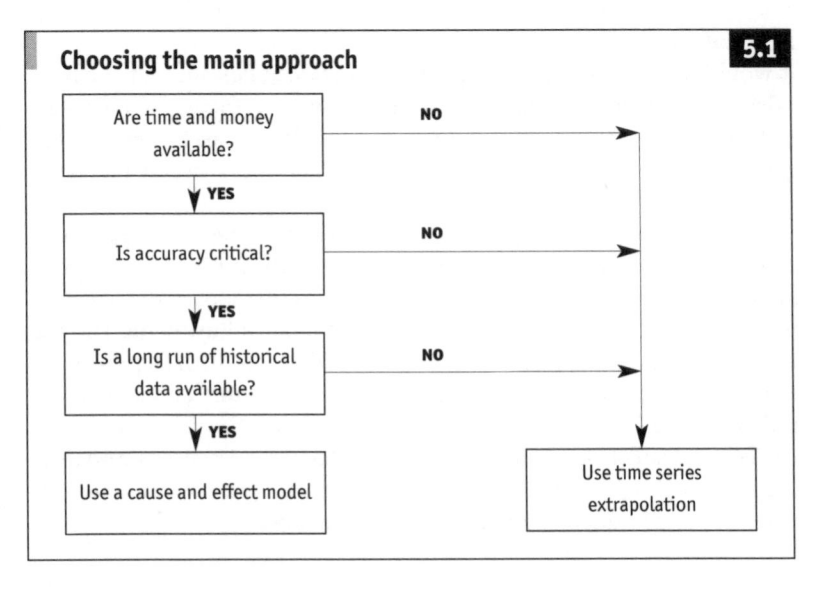

Choosing the main approach   5.1

- Are time and money available? — **NO** →
- ↓ **YES**
- Is accuracy critical? — **NO** →
- ↓ **YES**
- Is a long run of historical data available? — **NO** →
- ↓ **YES**
- Use a cause and effect model
- Use time series extrapolation

according to whether it measures value, volume, or prices. The number of tons of sulphuric acid shipped is a volume indicator. The income from those sales is a value indicator. The difference between the two is a price indicator. The relationship is not difficult to grasp: for any period, volume times price equals value.

It is sometimes necessary to measure volumes in money terms. For example, in 2000 a small manufacturer of agricultural equipment produced ten tractors which sold for $30,000 each and ten hand shovels worth $20 each. It is not particularly helpful to summarise total output as 20 items or 20 tons of engineering. It is meaningful to say that total output was valued at $300,200. This is apparently a value indicator. But if output in future years is also valued in the prices charged in 2000, regardless of actual market prices, the time series so created will record the total volume of production measured in 2000 prices.

Current and constant prices
When current (actual) selling prices are used, a series is a value or current price series. Current prices are also known as nominal prices. If prices applicable in some base period are used, 2000 in the example above, the result is a volume or constant price series. Such a series is said to show changes in "real" terms even though the result is never encountered in real life.

Inflation
It would not be necessary to distinguish between current and constant prices if prices did not change over time. However, there is evidence of inflation in the earliest financial records. Inflation is an increase in the general price level. It is tricky, because not all prices rise at the same pace.

Variations in the rate of inflation can obscure other factors such as changes in the underlying demand for a product. It is often easiest to analyse and forecast each item in volume terms (perhaps in the constant prices of today's money), and make individual adjustments for the expected course of inflation. For composite forecasts, such as profits, do not just add up all the components and then make one adjustment for inflation: wages and retail prices almost always rise at different rates.

Deflation. Statisticians frequently collect data in current prices and then deflate into constant prices. For this reason, the price index used for adjustment is frequently known as a price deflator.

The composition of a time series

Time series are interesting. Take a look at the graph at the top of Figure 5.2. This might show any one of a whole range of variables. It does not matter what is shown, the question is how can this seemingly random squiggle be analysed? In fact, like most value or volume time series, the squiggle comprises four component parts which can be identified separately.

The trend. This is the general, long-run path of the data. In Figure 5.2 there is a steady if gentle rise in the trend, perhaps due to growing demand for the product illustrated.

The cycle. Business data frequently contain one or more cyclical component. These may reflect industry or product life cycles or the economy-wide business cycle. Most economies exhibit a rhythmic recession-depression-recovery-boom pattern which lasts typically for four or five years. Demand for many products will rise during times of economic boom and fall during slumps, with knock-on effects on production, employment and so on. Cycles are rarely measured in months and may stretch to 50 or more years. Figure 5.2 shows an artificially smooth cycle of about five years' duration.

The seasonal pattern. Seasonality is a very short-term pattern which repeats every 12 months (eg, increased demand for Christmas cards in December). It is one of the most important elements to identify when analysing and forecasting short-term developments. Seasonality is not relevant if data are annual totals which by definition contain a complete seasonal pattern.

The residual. It is rarely possible to explain every factor which affects a time series. When the trends, seasonality and cycles have been identified, the remaining (ie, residual) unexplained random influence is all that is left.

Figure 5.2 is worth studying as it illustrates how one set of numbers can yield a wealth of information.

The decomposition of time series

The forecasting trick is to decompose a series into the trend, cycles, seasonality and residual, extend each into the future, and then recombine them into one again. The component parts may be thought of as adding or multiplying together to form the overall series (call this y). In shorthand, either $y = t + c + s + r$ or $y = t \times c \times s \times r$.

The second, multiplicative, model is more appropriate for most business situations. For example, a retailer sells $1,000 worth of greetings

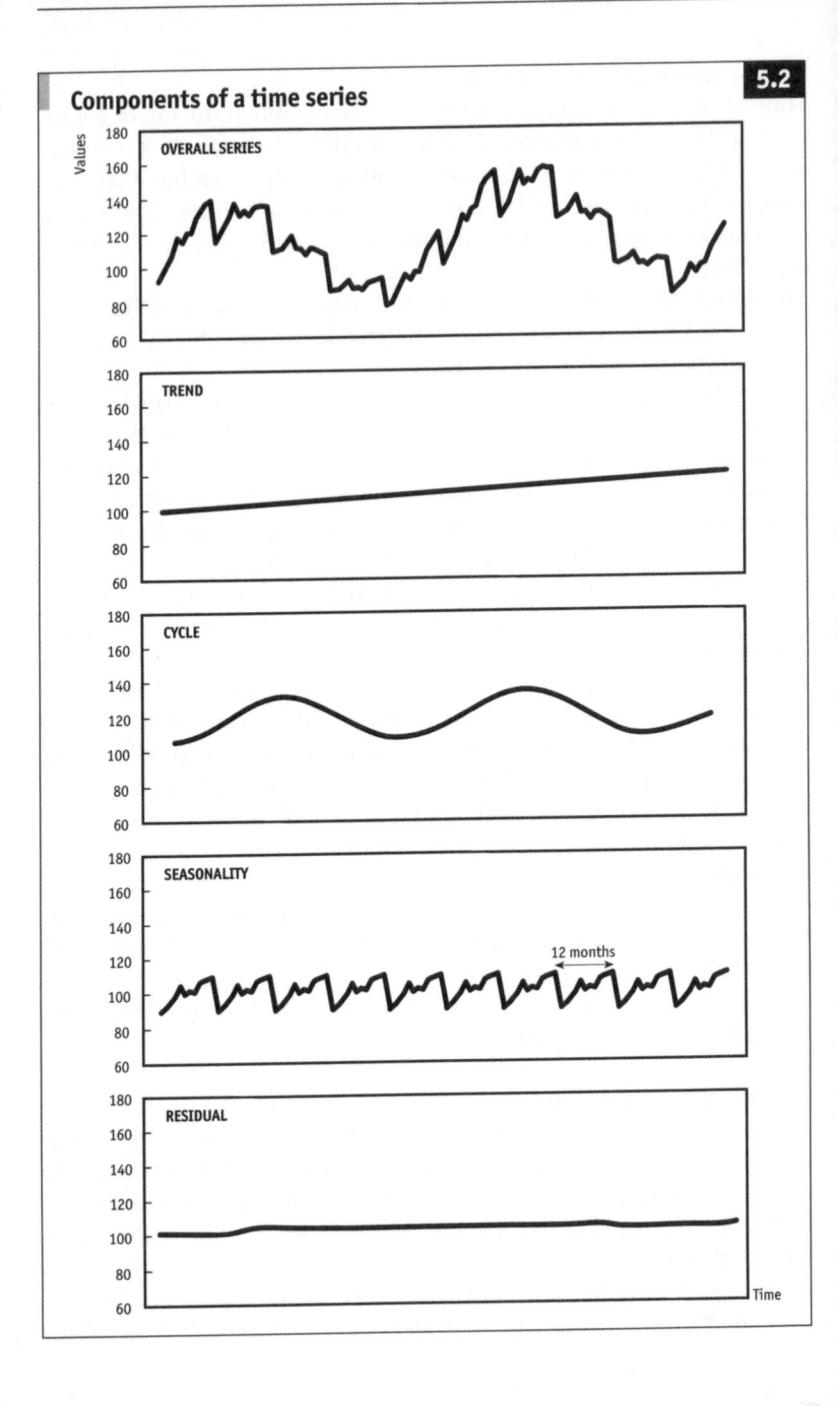

5.2

Components of a time series

cards every month except December when seasonal influences push card sales to $1,100. Is this seasonal variation additive (base amount + $100) or multiplicative (base amount × 110%)? The answer is probably the latter, since if regular monthly sales suddenly doubled, perhaps due to higher demand or inflation, it would be more realistic to expect December's sales to be $2,000 × 110% = $2,200 rather than $2,000 + $100 = $2,100.

Which components? The general starting point is to consider which influences (trend, seasonality, cycles or inflation) are relevant for the forecast. The trend is always important, but you do not need to waste time looking for seasonal patterns in annual totals or in data which are known to be free of seasonality, such as fixed monthly rental or loan payments. Similarly, you may know that cycles are not present or you may consider them to be unimportant for a short-term forecast. Inflation is not relevant if the forecast is not affected by prices, perhaps because it relates to the output of a particular machine which is determined endogenously (by internal factors such as technical constraints) rather than exogenously (by external influences such as consumer demand).

The procedure in outline. The procedure for full analysis of a time series is as follows. It may appear complex, but it is relatively straightforward in practice.

- Convert the original current series into constant price terms. For each period, divide the original value by the price deflator. There are now three series: the original data, the price deflator and the deflated series. This may be written in shorthand as follows: $Y(T \times C \times S \times R) = Y(P \times T \times C \times S \times R) \div Y(P)$, where $Y(P)$ is the price deflator.
- Identify the trend using moving averages, regression, or a more complex trend-fitting model. Create a (fourth) series which records this trend. For each period, divide the trend into the original data to create a (fifth) series which is a detrended version of the original: $Y(C \times S \times R) = Y(T \times C \times S \times R) \div Y(T)$.
- Identify the cycle. The simple approach is to assume that annual data contain only the trend and a cycle. Create a (sixth) series which represents the cycle by dividing annual data by the trend: $Y(C) = Y(T \times C) \div Y(T)$. Then convert the cycle into monthly or quarterly values and create a (seventh) series which is free of the cycle: $Y(S \times R) = Y(C \times S \times R) \div Y(C)$.

◪ Identify the seasonal pattern in the data (series eight) and divide this into the deflated, detrended, decycled series to identify the residual (series 9): $Y(R) = Y(S \times R) \div Y(S)$.

◪ Extend the indivdual components, and put them back together: $Y (P \times T \times C \times S \times R) = Y(P) \times Y(T) \times Y(C) \times Y(S) \times Y(R)$.

Trends

It is nearly always helpful to draw a graph of the series under review. In some cases, a trend can be drawn by eye and extrapolated with a ruler. This idea is discussed in the section on regression analysis, pages 104ff. This section reviews a few general techniques for smoothing out fluctuations and erratic influences.

Moving averages

A manager considering the typical level of mid-year sales might take the average of figures for May, June and July to iron out erratic influences. A sequence of such averages is called a moving average. (See Table 5.1 and Figure 5.3.) Moving averages are easy to calculate, especially when using a PC spreadsheet, so they are a quick way of identifying broad trends in a time series. Moving averages are important analytical tools, although they can be projected only by eye.

Calculating a moving average. A three-month moving average is found by adding values for three consecutive months and dividing the result by 3. Repeating this for each set of three months produces a new series of numbers. Table 5.1 shows how this is done.

Length. Moving averages can cover any number of periods. Figure 5.3 shows three-month and 12-month moving averages. An analyst smoothing out daily stock market data might take a 180-day moving average. The greater the number of periods, the smoother the curve traced by a

Table 5.1 **Calculating a three-month moving average**

	Raw stock index	3-month moving average	Calculations
Jan	183.5	–	
Feb	185.0	185.0	= (183.5 + 185.0 + 186.6) ÷ 3
Mar	186.6	186.2	= (185.0 + 186.6 + 186.9) ÷ 3
Apr	186.9	189.4	= (186.6 + 186.9 + 194.7) ÷ 3
May	194.7	193.1	= (186.9 + 194.7 + 197.8) ÷ 3
Jun	197.8	–	

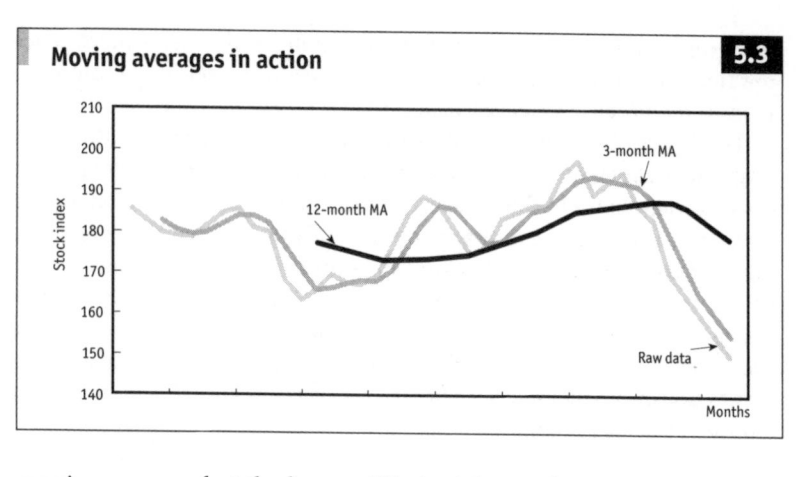

Moving averages in action 5.3

moving average, but the less rapidly it picks up changes in direction in the raw data.

Centred averages. Moving averages should be centred. For the three months January, February and March, the average relates to February. There are two difficulties.

1 A moving average for an even number of periods does not have an exact centre. There are two ways round this. The technically correct approach (taking an odd number of observations and weighting the middle ones, as in Table 5.4) is often more trouble than it is worth. The easy alternative is to leave the average off-centre.

2 Centring a moving average results in a number of periods at each end without a value (see Table 5.1). This may be acceptable for looking at the past, but it is not very helpful for forecasting.

Off-centre averages. In practice, many people place moving averages at the ends of each sequence of periods. The advantage is that this instantly reveals where the series is in relation to the trend. Stockmarket pundits frequently use this as a guide to the immediate future. If stock prices are below the moving average (as in Figure 5.3) they are expected to rise. If they are above the average they are expected to fall. This holds good only if the moving average is sensitive enough to the trend and the trend is not secretly undergoing a change in direction. This is always the problem when predicting the future from the past.

Moving averages for forecasting. It was emphasised that each moving average rightly belongs in the middle of the sequence of data from which it is calculated. For example, in Table 5.1, the final moving average of 193.1 should be centred against the month of May. The previous paragraph indicated that stockmarket pundits and others might place the figure against June. A simplistic approach to forecasting would use 193.1 as the predicted figure for July.

This works best with trendless data without a seasonal pattern. For example, a canteen serving a fixed population of 500 company employees experiences wide fluctuations in the number of meals served each day, but over time the average number of meals is neither increasing nor decreasing (demand is stationary). A moving average might be a satisfactory way of predicting tomorrow's demand for meals. If there is a seasonal pattern in the data (perhaps more employees stay on the premises for lunch during cold winter months), the moving average would perform better as a forecasting tool if the seasonal element is first removed (see pages 98ff).

Weighted moving averages

For forecasting, the latest values in a series can be given more importance by weighting the moving average. This is subjective, but it tends to make predictions more accurate.

The weights are picked from the air, pretty much on instinct. For simplicity, make them total 1. Say you want to give 70% importance to last month's sales, 20% to the month before and 10% to the month before that. Weights of 0.7, 0.2 and 0.1 are put into the calculation. Using the figures from the final three months in Table 5.1, the arithmetic is (186.9 × 0.1) + (194.7 × 0.2) + (197.8 × 0.7) = 18.69 + 38.94 + 138.46 = 196.1. Note that there is no need for any division since the weights sum to 1.

The snag is that if there is a seasonal pattern in the data weighting will amplify the effects of the latest seasonal influences. As with simple moving averages, weighted moving averages work best with data which are free of seasonal influences.

Exponential smoothing: the easy way to project a trend

Exponential smoothing is a delightful way to use weighting. Make a forecast for period one and pick a single weight, say 0.2. Then sit back and watch. These two figures alone will generate a one-period forecast for ever more. Table 5.2 shows how the arithmetic is performed.

Choosing the weight. With some manipulation, it can be shown that this exponential smoothing is a simplified form of a geometric progression. The weight, which must be between 0 and 1, gives declining importance to a whole sequence of earlier periods. When the weight is near 0.1, it takes around 20 previous periods into account. When it is 0.5, it

Table 5.2 **Exponential smoothing**

	Raw stock index	Forecast	Calculations
Jan	183.5	–	
Feb	185.0	184.5	= seed forecast
Mar	186.6	184.7	$= 184.5 + [0.4 \times (185.0 - 184.5)]$
Apr	186.9	185.5	$= 184.7 + [0.4 \times (186.6 - 184.7)]$
May	194.7	186.1	$= 185.5 + [0.4 \times (186.9 - 185.5)]$
Jun	197.8	189.5	$= 186.1 + [0.4 \times (194.7 - 186.1)]$
Jul		192.8	$= 189.5 + [0.4 \times (197.8 - 189.5)]$

Calculations: the forecast for July

= forecast for June + [weight × (outturn for June − forecast for June)]

= 189.5   + $[0.4 \times (197.8 - 189.5)]$

uses only the last three periods. A value of between 0.1 and 0.3 usually works well. Establish the best figure by trial and error (produce "forecasts" for past periods using different weights and identify the weight which gives the best "forecast" – see Forecast monitoring and review, pages 113ff). A weight greater than about 0.5 suggests that the data are jumping around so much that other forecasting techniques are required, perhaps seasonal adjustment.

Table 5.2 and Figure 5.4 show exponential smoothing used to predict

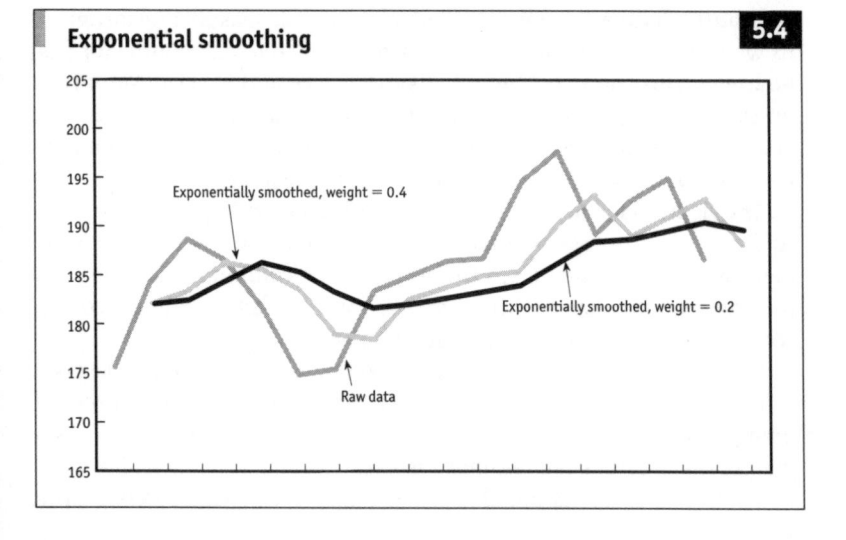

Exponential smoothing   5.4

Exponentially smoothed, weight = 0.4

Exponentially smoothed, weight = 0.2

Raw data

stockmarket prices. You would not do this in practice, since future stock prices depend more on investor expectations than on historical developments. The illustration is useful, however, because it shows the way that an exponentially smoothed series reacts to changes in the underlying data.

Choosing the initial forecast. The first seed forecast can be found by various guesstimates, such as from a simple moving average.

Using exponential smoothing. Exponential smoothing is used successfully by many business organisations. It is simple and easy to automate with a PC. As always, note that it uses only historical data and responds slowly to changes in trend. Some computer packages include advanced smoothing techniques which take better account of trends and the other components of time series, which are discussed below.

Seasonal adjustment

Seasonality needs little explanation. Demand for swimwear and gardening tools is higher during the summer. More than half of all the perfume sold worldwide every year is bought in the four days before Christmas.

Whenever there is a seasonal factor it is important to recognise it and compare sales in the seasonally affected period with the same period the previous year. Sales in November and December may be streets ahead of sales in September and October but what is more relevant is how they compare with November and December last year.

12-month changes. A quick way of coping with seasonal distortions is to work with changes (absolute or percentage) over 12 months (or four quarters). This compares broadly like with like and is very useful for a quick check on (non-seasonal) trends. Table 5.3 shows this procedure in action, and shows how taking several months together (essentially, using a moving average) smooths out erratic fluctuations.

The figures in column B are an index of sales, perhaps by a gift shop. At first glance it appears that sales in January 2003 were poor, since there was a 4% decline from December 2002 (column C). It seems that this interpretation is confirmed because the 4% fall in January 2003 is worse than the 0.2% decline in January 2002.

However, the percentage change over 12 months (column E) gives some encouragement. It indicates that the trend in sales is upward, though growth over the 12 months to January 2003 (6.2%) was slacker than in the previous few months (around 10%).

Column G smooths out short-term erratic influences by comparing

Table 5.3 **Analysing seasonality in a short run of data**

A	B	C	D	E	F	G
			Percentage changes		Total of	% change
	Sales	1	1 month	12	3 months'	3 month on
	index	month	annualised	months	sales	3 month
Oct 2001	119.2					
Nov	120.2	0.8				
Dec	121.8	1.3			361.2	
Jan 2002	121.6	−0.2			363.6	
Oct	130.7	−0.5	−6.2	9.6		
Nov	133.4	2.1	27.8	11.0		
Dec	134.4	0.7	9.3	10.3	398.5	10.3
Jan 2003	129.2	−3.9	−37.7	6.2	397.0	9.2

Calculations for January 2003:

Column C: $[(129.2 \div 134.4) - 1] \times 100 = -3.9\%$

Column D: $[(129.2 \div 134.4)^{12} - 1] \times 100 = -37.7\%$.

The figures in column D show the percentage change which would result if the change over one month was repeated (compounded) for 12 consecutive months. Such annualising is popular in the USA, but note how it amplifies erratic influences. The arithmetic is discussed in Percentages and proportions, pages 12ff.

Column E: $[(129.2 \div 121.6) - 1] \times 100 = 6.2\%$

Column F: $133.4 + 134.4 + 129.2 = 397.0$

Column G: $[(397.0 \div 363.6) - 1] \times 100 = 9.2\%$

sales in the latest three months with sales in the same three months a year earlier. This suggests that the fall in January was not as severe as it appeared at first glance, with the 12-month growth rate remaining at close to 10%.

This final interpretation is the correct one. Indeed, a full run of figures would emphasise that sales fell in January only because this was a correction to an exceptionally steep rise in the earlier few months.

The seasonal pattern. A better way to proceed if you have several years' data is to identify the quarterly or monthly pattern for a typical year. The pattern can be summarised in a set of four or 12 index numbers which describe the path of sales, output and so on when there is no trend, no cycle and no random influence. This might not suggest a particularly demanding environment for business. The seasonal pattern, however, is crucial for business planning, since it affects ordering, staffing needs, cash flow, and so on.

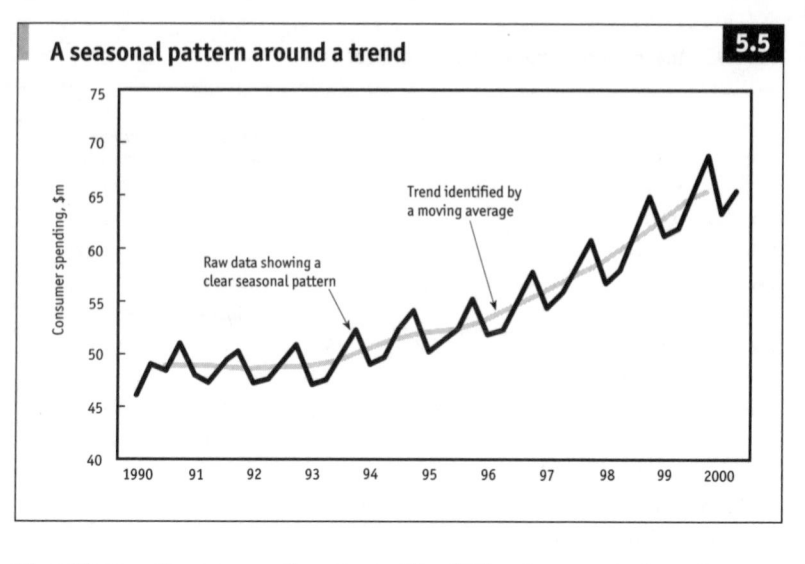

A seasonal pattern around a trend 5.5

Identifying the seasonal pattern. The following steps show how to identify the seasonal pattern in a set of quarterly figures. (For monthly data, or any other frequency, replace four with 12 and quarter with month, etc.)

1 Identify the trend in the quarterly series. Figure 5.5 shows the seasonal pattern in relation to a trend identified with a moving average (regression analysis might be used instead). The moving average must be centred correctly. The calculations for Figure 5.5 are illustrated in Table 5.4, box 1.

2 Divide each original value by the corresponding moving average value to create a third series (column D in Table 5.4.)

3 Collect together all the first quarter figures for this third series. Discard any extreme outliers (the most random influences) and find the average (arithmetic mean) of the remaining first quarter figures (Table 5.4, box 2).

4 Repeat this for the other three quarters.

5 Add together the four values resulting from steps 3 and 4.

6 Convert the four values resulting from steps 3 and 4 into index form by multiplying each value by (4 ÷ total from step 5).

Interpreting the seasonal pattern. The end product of the six steps just described is a set of four figures (which themselves add up to a total of 4). Call them seasonal adjustment factors. They are interesting because they show the seasonal pattern hidden in the original data. For example, in Table 5.4, the seasonal adjustment factor for the first quarter is 0.970.

This implies that in a typical first quarter, sales are 97% of the average quarterly level. The pattern is especially useful for planning when it is derived from monthly data.

Seasonal adjustment. Having found a set of seasonal adjustment factors, the original data series can be deseasonalised by dividing all 1Qtr figures by the 1Qtr adjustment factor, dividing all 2Qtr figures by the 2Qtr adjustment factor, and so on (Table 5.4, column F).

The seasonally adjusted series might be projected (by fitting a trend line with regression analysis or by exponential smoothing). The seasonality is put back into the projected series by multiplication. For example, if in Table 5.4 the projected seasonally adjusted sales figure for the first quarter of 2003 is $71,000, the unadjusted figure (ie, the forecast of actual sales) would be 71,000 × 0.970 = $68,900.

Advantages and pitfalls of seasonal adjustment. Seasonality is identified using simple numerical methods and applied mechanically. It

Table 5.4 **Full seasonal adjustment**

A	B		C	D	E	F
				Detrended	Seasonal	Seasonally
	Sales		Moving	series	adjustment	adjusted sales
Period	$000s		average	B ÷ C	factors	B ÷ E
2000 1Qtr	56.6		58.9	0.962	0.970	58.3
2Qtr	58.0		59.8	0.970	0.973	59.6
3Qtr	61.7		60.9	1.013	1.011	61.0
4Qtr	65.0		62.0	1.049	1.046	62.1
2001 1Qtr	61.1		63.0	0.970	0.970	63.0
2Qtr	62.0		64.0	0.969	0.973	63.7
3Qtr	65.9		64.7	1.018	1.011	65.2
4Qtr	68.8	→	65.4	1.052	1.046	65.8
2002 1Qtr	63.3				0.970	
2Qtr	65.5				0.973	
				↓	↑	

Box 1 The moving average
Each entry in Column C is calculated in the same way as this example for 2001 4Qtr:

$$(62.0 \times 1) = \quad 62.0$$
$$+(65.9 \times 2) = \quad 131.8$$
$$+(68.8 \times 2) = \quad 137.6$$
$$+(63.3 \times 2) = \quad 126.6$$
$$+(65.5 \times 1) = \quad \underline{65.5}$$
$$= 523.5 \div 8 = 65.4$$

Division by 8 reflects the fact that the weights sum to this amount.

Box 2 The seasonal adjustment factors
The figures in column E are the four seasonal factors found by averaging the figures in column D. The calculations use a longer run of data to more decimal places than shown above.

Average of Q1s	0.97015 × 0.99990 =	0.97006
Average of Q2s	0.97325 × 0.99990 =	0.97316
Average of Q3s	1.01102 × 0.99990 =	1.01092
Average of Q4s	1.04597 × 0.99990 =	1.04587
Total	4.00039	4.00000
Divide into 4	0.99990	

improves historical analysis and forecasting without requiring any great judgmental abilities. However, there are two points to watch.

1 Rather more advanced techniques are required to trap tricky variable influences such as the number of shopping days in a given month and the date of Easter, all of which differ each year.

2 Seasonal adjustment is tied to a particular sequence of months. Adding 12 (seasonally adjusted) values to obtain an annual total requires care. For example, fiscal year totals should not be calculated from a seasonally adjusted series constrained to (ie, based on) calendar years. Seasonal adjustment factors which sum to 12 over a calendar year may not sum to 12 for the fiscal year.

Cycles

Cycles are longer-term influences which are difficult to detect without very long runs of data. If a cycle has a ten-year duration, you need to analyse perhaps 50 years of data before you can make a sensible estimate of the length and magnitude of the average cycle. Clearly this is not always feasible, so it is often necessary to ignore the cyclical component when making short-term forecasts.

Nevertheless, it is important to identify turning points. An easy solution is to look for links between the independent variable (such as sales) and a leading indicator of cyclical trends (such as consumers' income). (See Leads and lags, page 103.)

There might be more than one cycle, perhaps a five-year business cycle and a ten-year product cycle. Cycles can usually be spotted by drawing a graph of annual figures and observing the way the curve fluctuates around the trend.

Residuals

When a time series has been analysed almost to extinction, the only remaining element is the residual component. Always draw a graph of the residuals to make sure that they are randomly scattered. A regular pattern indicates that the analysis can be improved. For example, a large residual value recurring every January would suggest that the seasonal adjustment process had not taken account of all the seasonal influences (see also Regression residuals, pages 112ff).

Analysis of time series is not a perfect science. Time, money, or mathematical constraints may force you to terminate your analysis even though there is still a fairly regular residual element. In such cases, an educated judgment about the residuals may improve the forecast. For example, if your analysis leaves a residual of around $100 per month for recent historical data, you might project your identified trend and add the "unexplained" $100 to each figure to obtain the working forecast.

Cause and effect

A thoughtful approach to forecasting is to base predictions on established relationships. For example, the managers of a retail chain conclude that their sales depend directly on the local population (see Figure 5.6) according to the following relationship: sales = 2 + (4 × population), where both sales and population are measured in millions. From this, the managers predict that if they open a new outlet in an area with a population of 2.5m, their annual sales will be 2 + (4 × 2.5) = €12m.

Leads and lags

A reduction in income tax rates leaves more money in consumers' bank accounts which may well be spent on cars or clothes. The question is when? A tax cut in April may not make shoppers feel wealthy enough to spend more until July. Alternatively, a much trumpeted fiscal change might encourage consumers to spend in February in anticipation of higher pay packets to come. Thus, spending may lag or lead changes in income tax. Deciding in advance what will happen requires good judgment.

Leading indicators. There are many examples of leading and lagging relationships. It is helpful to identify leading indicators which affect whatever is to be forecast. A retailing group might use average earnings, employment and interest rate changes to predict future sales. A wholesaler might use these plus government figures on high street sales; a manufacturer might add distributors' inventories-to-sales ratios, and so on. Governmental and private statistical agencies publish many useful basic time series. They also compile leading indicators which are composite indices of several series that help predict turning points.

Lagging. Quantifying a link between a leading indicator and the dependent variable is partly trial and error. Imagine drawing a graph of each indicator, then holding them up to the light one on top of the other and moving one sideways until the two curves coincide. Comparison of the time scales on the graphs will show by how many months one series lags the other. Regression analysis allows you to make the comparison using a numerical rather than a visual method.

For example, you might decide that sales lag interest rates by three

Published sources

Composite leading indicators are published by agencies such as the National Bureau for Economic Research in the US and the Office for National Statistics (ONS) in the UK.

There are a great many other published series which are valuable for analysis and forecasting. National government sources include US departments such as the Department of Commerce and the Federal Reserve. In the UK the ONS draws many series together in monthly publications such as *Economic Trends* and the *Monthly Digest of Statistics*. Every week *The Economist* publishes in its back pages economic and financial statistics for 15 developed and 25 emerging economies, and the euro area. These tables cover output, unemployment, inflation, trade, exchange rates, interest rates and stockmarkets in a handy, comparative format.

More detailed collections of monthly statistics covering many countries are published by international organisations. Key sources include *Main Economic Indicators* for 29 member countries of the Organisation for Economic Co-operation and Development (OECD); *International Financial Statistics* for all International Monetary Fund (IMF) member countries published by the IMF; and various reports produced by the EU.

Valuable advance indicators are prepared by business organisations such as the Conference Board and the National Association of Purchasing Management in the US and the surveys conducted by the Confederation of British Industry in the UK.

Large swathes of statistical information are now available on national statistics offices' and international organisations' websites; see www.NumbersGuide.com or, for example, www.bea.doc.gov, www.statistics.gov.uk, www.bls.gov, www.imf.org, europa.eu.int.

months. In other words, sales in, say, June are influenced by the level of interest rates during the previous March. You would move the interest rate series so that March interest rates were paired with June sales (and likewise for all other periods) and use regression to establish a precise relationship. Then at the end of, say, July when you know the actual level of interest rates during that month, you can predict your sales for the coming October.

Identifying relationships with regression analysis

Analysis and forecasting (extrapolation or causal) can be improved by identifying the precise relationship between one set of numbers and another. There is a rather neat numerical approach which was pioneered by a statistician who was interested in the way that tall parents' sons regressed towards the average height of the population. Consequently, the technique is stuck with the awkward name of regression analysis.

Regression identifies the equation of the line of best fit which links two sets of data (as in Figure 5.6) and reveals the strength of the relationship. One set of data may be time, in the form of years (eg, 2000, 2001, 2002 ...) or arbitrary numbers (eg, January 2000 = 1, February = 2, ... January 2001 = 13, and so on).

The general approach is as follows:

- line up the two series to be regressed;
- find the equation which describes the relationship; and
- test the strength of the relationship (the correlation) between the two series.

Essentially regression analysis fits a line which minimises the average vertical distance between the line of best fit and each scattered point. Some points will be above the line (positive deviation) others will be below (negative deviation). The negative signs are disposed of by taking the squares of the deviations (two negative numbers multiplied together make a positive number), so the technique is also known as least squares regression.

It is straightforward, if tedious, to perform the necessary calculations by hand. In practice, calculators and PC spreadsheets do the work quickly and easily. With a PC spreadsheet, select data regression from the menu, identify the sets of data to be regressed and a spare area of the worksheet where the results will be reported, and tell the PC to proceed. The results will appear almost instantaneously. Some calculators will perform regression. You have to indicate that you want to do regression, key in the pairs of values, tell the calculator that you have finished the data entry, and press another sequence of buttons to display the results.

PC spreadsheets are recommended. They retain and display all the data for error-checking and further analysis, and they generally provide far more information about the relationships.

The relationship
Simple regression identifies the relationship between two variables.

Identifying a straight line 5.6

The managers of a retail chain notice a link between sales at their outlets and the size of the local population. A scatter graph of these two variables (sales, population) is shown below. The straight line of best fit was fitted by regression analysis. This line, and every other straight line, has a simple label which identifies it as unique from other lines. The label (or equation) is most conveniently written as $y = a + (b \times x)$: where x and y are variables (population and sales in this case) and a and b are constants. The constants a and b are very easy to find.

The first, a, is no more than the point at which the line hits (intercepts) the vertical axis (where $x = 0$). Read this from the graph: 2 in this example. The second constant, b, is the slope of the line. This can be found from any pair of co-ordinates. One pair is already known, where the line crosses the y-axis (0, 2). Call this (x1, y1) in general. Pick another pair from anywhere on the line, say, (2, 10) or (x2, y2) in general . The slope is

$$(10 - 2) \div (2 - 0) = 8 \div 2 = 4 \text{ or, in general } (y2 - y1) \div (x2 - x1).$$

If the slope b is negative, it indicates that the line dips down towards the right. If b is zero, the line is horizontal. A positive b indicates an upwardly sloping line. If b is infinitely large, the line is vertical. Most planners hope for a huge b when forecasting sales.

The equation of the line below is $y = 2 + (4 \times x)$ or sales $= 2 + (4 \times \text{population})$. It might seem odd that this line implies that sales will be €2m even when the population is zero. This reflects the imperfection of the model linking sales and population. Such flaws may not be obvious, but always look for them.

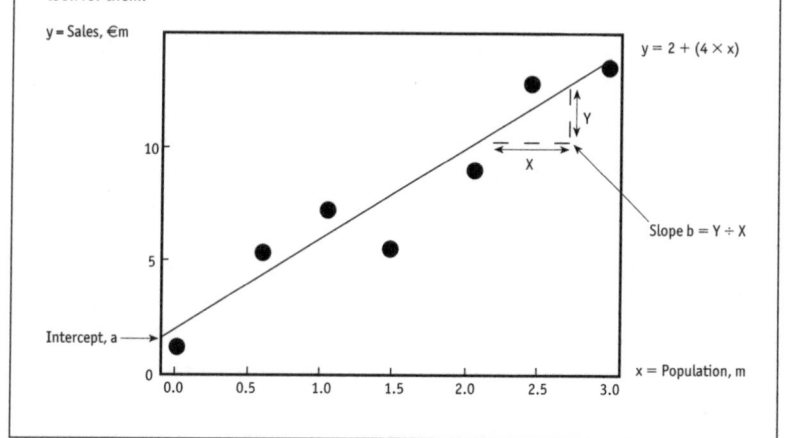

1 The dependent variable or y variable is the item being forecast (such as sales). It is usually measured along the vertical axis of a graph.

2 The independent variable or x variable is the explanatory series (such as population or advertising expenditure). It is usually measured along the horizontal axis of a graph. Note that in forecasting exercises, particularly when regression is being used to identify a trend, the independent variable is often time.

The dependent and independent variables are linked by the equation of a straight line, $y = a + (b \times x)$, where a and b are constants; b is sometimes called the x coefficient (see Figure 5.6). The first information to come from regression analysis is the numerical values for a and b. These allow y to be estimated for any value of x. For example, in Figure 5.6 a and b are 2 and 4 respectively. The equation is $y = 2 + (4 \times x)$, so for population $x = 4.5$, sales $y = 2 + (4 \times 4.5) = €20$ million.

The strength of the relationship

Regression analysis provides some important indicators of how the identified relationship works in practice.

The coefficient of correlation, better known as r, identifies the correlation (the strength of the relationship) between the two sets of data represented by x and y. This coefficient will always be between -1 and $+1$. If $r = 0$, there is no linear correlation between the two series. If $r = 1$ or $r = -1$ there is perfect linear correlation. The sign can be ignored since it indicates no more than the slope of the line, which is already evident from the calculated value of b.

In Figure 5.6, $r = 0.93$. This indicates a fairly strong correlation between x and y (population and sales). Figure 5.7 shows some other scatters and the associated values of r. Note that the value $r = 0$ for the top right graph indicates that no linear relationship exists even though there is a curvilinear link between the points.

The coefficient of determination, better known as $r^2 (= r \times r)$, indicates how much of the change in y is explained by x. This magic r^2 will always be between 0 and 1. In Figure 5.6, r^2 is 0.86. So 86% of sales is explained by the size of the local population. Note, though, that explained by does not mean caused by. There may be a close correlation between the annual value of Irish whiskey sales and the income of Irish priests, but are the two related to each other? More probably they are each linked to some third series, such as inflation.

Standard error. The standard error is the standard deviation of the estimate. Interpret it in exactly the same way as any other standard deviation (see pages 63–64). For example, if a regression equation predicts sales at €150m and the standard error is €3m, there is a 95% chance that sales will be between €144m and €156m (ie, €150 ± two standard errors, or €150 ± 6).

Non-linear relationships

Business situations are rarely explained by simple linear relationships; they more commonly plot as curved lines. For example, production or

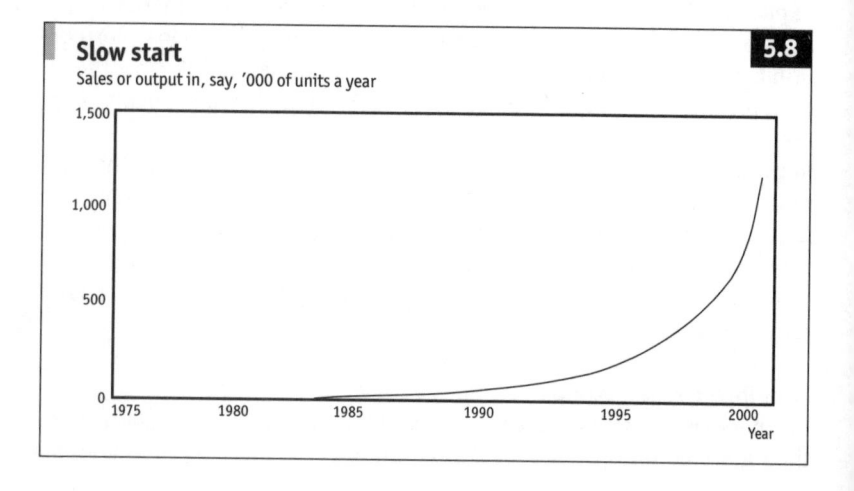

Sample correlation coefficients 5.7

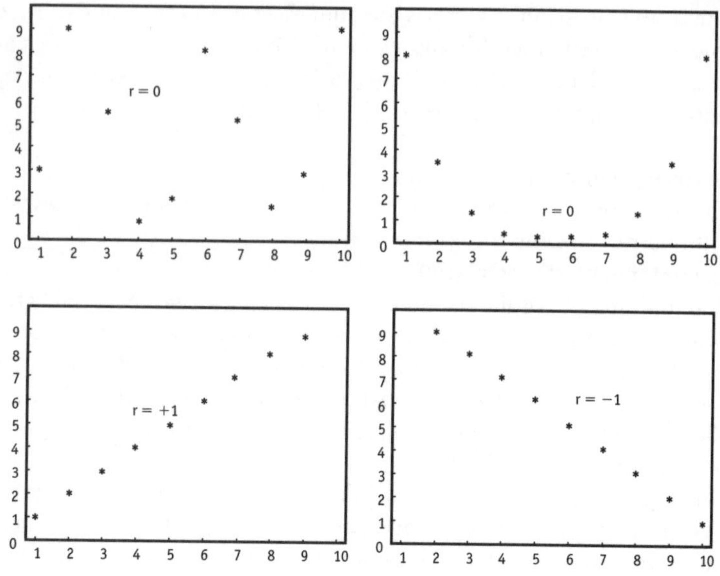

$r = 0$

$r = 0$

$r = +1$

$r = -1$

sales might start slowly before gathering pace (Figure 5.8), or they might taper off as the market becomes saturated (Figure 5.9). There are as many other possible curves and explanations as there are businesses and products.

Slow start 5.8

Sales or output in, say, '000 of units a year

Year

This might appear to present a problem, since most regression analysis is based around linear relationships. However, many curved lines can be transformed into straight lines which can be extended into the future using regression analysis, and then untransformed (as shown in Figure 5.10). There are many different ways of transforming curved lines into straight lines. For business purposes, the most useful transformation uses logarithms, which conveniently flatten out growth rates.

Logarithmic transformation. The sales figures in Figure 5.10, column B, plot as a curved line which is not amenable to linear regression. But the logarithms of the values plot as a straight line which can be identified with linear regression (column C). The straight line can be easily extended into the future (observations 9 and 10). Taking the anti-logarithms of the values in column C converts them back into the units of column B.

Logarithms are another name for the little number written in the air in powers (see page 18). For example, in $10^2 = 10 \times 10 = 100$, 2 is the logarithm of 100. In $10^3 = 10 \times 10 \times 10 = 1,000$, 3 is the logarithm of 1,000. Logarithms are a convenient way of handling large numbers and they convert growth rates (which are curvilinear) into simple arithmetic progressions (which are linear).

With calculators and PC spreadsheets use the keys/functions marked LOG or LN to find logarithms. For example, key 12.6 LOG to find that the logarithm of 12.6 is 1.10. To go back in the other direction (to take an antilog), use the 10^x or e^x (EXP) keys/functions. For example, keying 1.1 10^x will display 12.6.

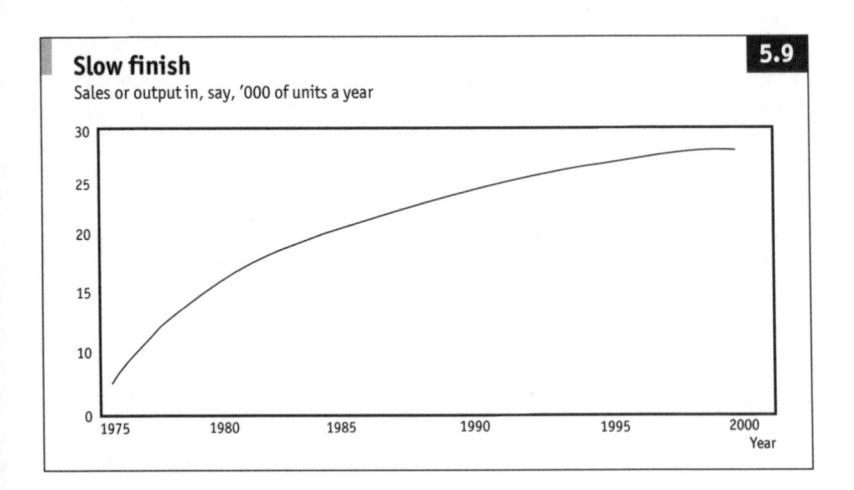

Slow finish   5.9

Sales or output in, say, '000 of units a year

Year

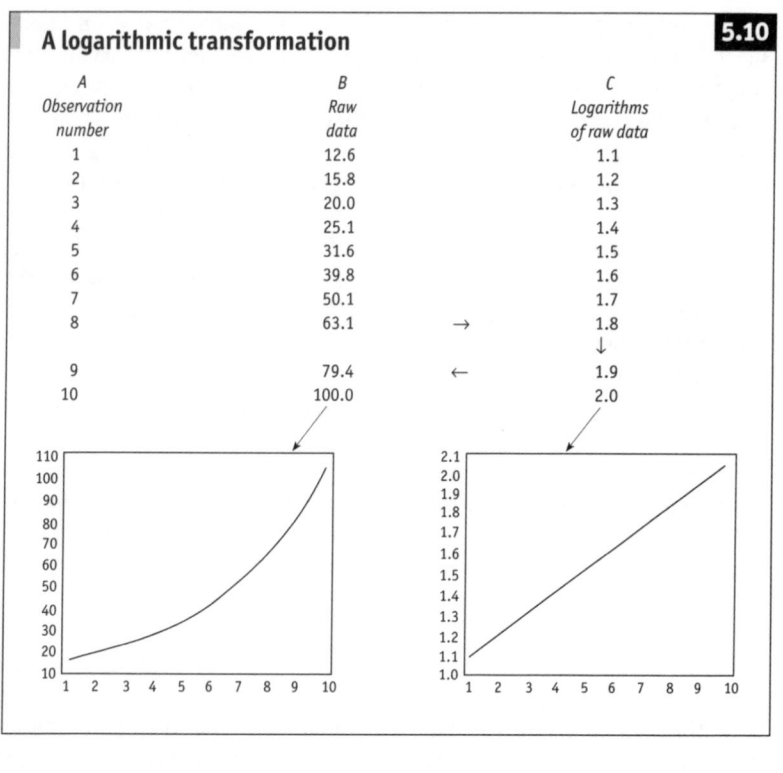

5.10

A logarithmic transformation

A Observation number	B Raw data		C Logarithms of raw data
1	12.6		1.1
2	15.8		1.2
3	20.0		1.3
4	25.1		1.4
5	31.6		1.5
6	39.8		1.6
7	50.1		1.7
8	63.1	→	1.8
			↓
9	79.4	←	1.9
10	100.0		2.0

Changes in direction

Figure 5.11 illustrates part of a typical product life cycle. Sales start slowly, gather pace as the product becomes popular, then slow down again as the market becomes saturated. Curves like this which change direction more than once cannot be flattened with transformations. There are some non-linear regression techniques for establishing such relationships, for which you need to use a computer or a specialist.

Super-bendy curves can be simplified by splitting them into two or three segments and treating each part as a straight line (or transforming each into a straight line). The problem, as always, is knowing when a change in direction is about to happen.

Multiple regression

So far the focus has been on two-variable relationships. This is clearly unrealistic in many instances. For example, the retail chain mentioned in Figure 5.6 might decide that its sales depend not just on the popula-

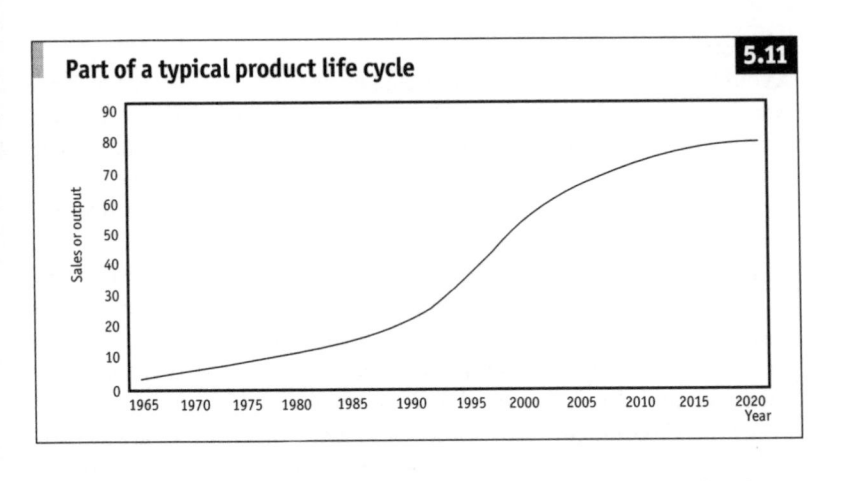

Part of a typical product life cycle   5.11

tion of the area in which it operates, but also on local income levels.

This three variable linkage could be plotted on a three-dimensional graph, with income rising up from the page. The scatter plot would look like small balls floating at different levels in a fish tank. The task is to poke in a line of best fit through all three dimensions. As the number of variables increases, the multi-dimensional problem becomes harder to visualise.

Fortunately, PC spreadsheets usually have multiple regression packages which handle the calculations with ease. Multiple regression works in exactly the same way as simple (two variable) regression.

The equation of the multiple regression line. The equation of a line linking several variables can be written as $y = b_0 + (b_1 \times x_1) + (b_2 \times x_2)$ with as many $+ (b \times x)$ terms as there are independent variables. Note that the intercept (called a in the equation of a straight line – see Figure 5.6) tends to be known as b_0 in multiple regression and is termed the constant. The other b values are known as partial (or net) regression coefficients.

r and r^2. As with two variables, the coefficients r and r^2 indicate the extent of the relationship between the independent variables taken as a group (the x values) and the dependent variable (y). In addition, multiple regression techniques also produce a sequence of partial correlation coefficients. Each indicates the relationship between the dependent variable (y) and one of the independent variables (say, x_1) when all the other independent variables (x_2, x_3, etc) are held constant. This is an economist's dream. It is one of the few occasions when they can examine things with all other influences held equal.

Choosing the variables

Avoid the blind approach of examining every series to find two or more which appear to be linked and then inventing a rationale for the link. This leads to spurious results. For example, finding a close relationship between the number of births and the number of storks in the Netherlands does not prove cause and effect. Instead, consider what relationships might exist, then use regression to test them.

Relationships between independent variables. A problem occurs when independent variables are highly correlated with each other. This is known as colinearity if there are two variables involved and multicolinearity if there are several. In such cases, the partial coefficients of regression and correlation need to be interpreted with extreme care.

A little forethought often indicates when colinearity might be present. For example, there is likely to be a strong correlation between prices for heating and transport fuels. Using both to predict, say, total energy demand could cause problems. The way around this is to:

- use only the more important of the two (perhaps heating costs);
- combine them into one series (eg, a weighted average of heating and transport fuel prices); or
- find a third series (eg, crude oil prices) to replace both.

t ratios. There is one t ratio for each independent variable. Very roughly, if t is between −2 and +2 the independent variable appears to have little effect in the equation and could be omitted. Use common sense to decide whether to include a variable you know to be important even if its t ratio suggests that it should be excluded.

Regression residuals

One test everyone should do with simple or multiple regression is to check the residuals. This is exactly the same concept as the review of residuals in time series analysis. The residuals should be random. If there is any pattern remaining then the analysis is not complete.

The residuals are simply the differences between each observed value and the corresponding value calculated using the regression equation. A graph of the residuals will often highlight any problem. In Figure 5.12 the pattern suggests that a straight line has been fitted to a curvilinear relationship; in this case a simple transformation will probably help.

Autocorrelation. Another problem is autocorrelation, where successive values of the dependent variable are affected by past values of itself.

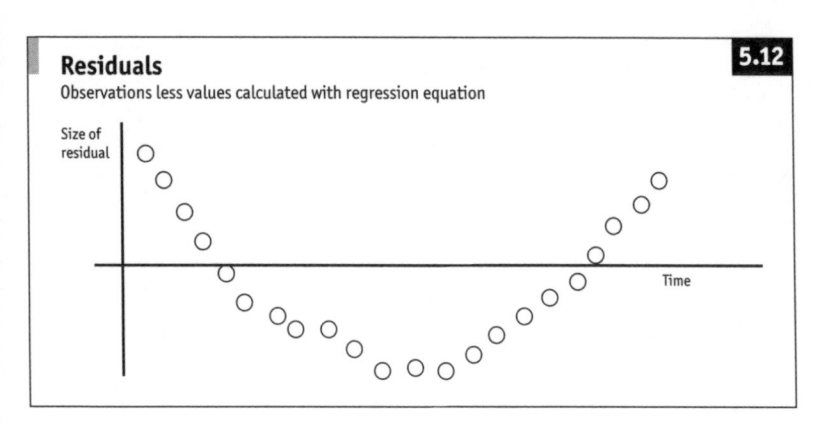

Residuals `5.12`

Observations less values calculated with regression equation

This is very common with time series. One example is where a department's annual budget is set each year at the previous year's figure plus 10%. Some PC regression programs produce the Durbin-Watson (DW) coefficient of serial correlation, which shows whether residuals are interconnected. DW lies between 0 and 4. A DW of 2 indicates the required absence of any serial correlation.

Forecast monitoring and review

It is obviously important to compare outturns (what actually happens) with projections. This provides a valuable insight which helps to improve the next forecast. The techniques for monitoring and review are also valuable as guides to selecting the best forecasting method to use in the first place. A good trick is to "forecast" for past periods, find the average error, and then see if a smaller error could have been obtained by using a different forecasting method, or a different exponential smoothing factor or regression equation. This retrospective review makes for better forecasting in the future.

For any one period, the error (outturn minus forecast) may be positive or negative. Dispose of the plus and minus signs by taking the absolute values (ie, ignore the signs) of the actual or percentage error, or by squaring the errors. The mean error over the forecasting period tells you how far out you were on average. For this reason, the measure of overall forecast accuracy is known by one of the following impressive terms.

- ◪ MAD: mean absolute deviation.
- ◪ MAPE: mean absolute percentage error.
- ◪ MSE: mean squared error.

Table 5.5 **Forecast monitoring and review**

A	B	C	D	E	F	G	H	I	J	K
1	125	123	2	2	1.6	4	2	2	2.0	1.00
2	130	133	−3	3	2.3	9	−1	5	−0.5	−0.20
3	131	136	−5	5	3.8	25	−6	10	−2.0	−0.60
4	134	135	−1	1	0.7	1	−7	11	−1.8	−0.64
5	138	136	2	2	1.4	4	−5	13	−1.0	−0.38
6	137	134	3	3	2.2	9	−2	16	−0.3	−0.13
7	139	136	3	3	2.2	9	1	19	0.1	0.05
8	140	142	−2	2	1.4	4	−1	21	−0.1	−0.05
9	141	140	1	1	0.7	1	0	22	0.0	0.00
TOTALS			0	22	16.4	66				
TOTALS ÷ 9			0.0	2.4	1.8	7.3				
				↑	↑	↑				
				MAD	MAPE	MSE				

Note: B: Outturn; C: Forecast; D: Error B–C; E: Absolute error |D|; F: Absolute % error E ÷ B; G: Squared error D × D; H: Running total of error D; I: Running total of absolute error E; J: Bias H ÷ A; K: Tracking H ÷ I.

Table 5.5 compares a sequence of forecast values with the actual outturns. Column D indicates that the forecasting errors cancel out overall, which does not reveal very much about the average error. Column E reveals that the average error, ignoring sign (ie, MAD), is $2.4. Column F shows that the average forecast, ignoring sign, was 1.8% out (MAPE). Column G puts the MSE at $7.3². By themselves, these numbers give some feel about the precision of the forecast. They become much more useful when comparing two or more different forecast methods to see which produces the smallest error. Which you choose is largely a matter of experience and personal preference. MAPE is a good general purpose measure since percentages are generally familiar and meaningful. On the other hand, if you can get used to the squared units of MSE, this measure may be better since it penalises large errors. One large error is generally more disastrous than several small ones.

Bias and tracking. MAD, MAPE and MSE allow you to review the accuracy of a forecast in retrospect. Bias and tracking provide useful running checks.

Bias indicates the average accuracy of the forecasting process to date (total error ÷ number of periods). In this example (Table 5.5, column I), the large number of negative values indicate a tendency for the outturns to be below the forecast value.

The tracking (bias ÷ MAD for the forecast to date) provides a running check on forecasting errors. It will alternate around zero if the forecasting is accurate. The steady (negative) growth in the early forecast period in Table 5.5 column J throws up a warning signal. Once you have run a few trial forecasts, you will develop a feel for the acceptable size of the tracking. Then, if a forecast with an unacceptably large tracking error arrives on your desk (ie, if tracking exceeds some predetermined trigger), you know it is time to review the forecasting method.

Delphi method

There is a well-documented judgmental forecasting process known as the Delphi method. Each member of a group of respondents (branch managers, heads of department, etc) completes a questionnaire outlining their expectations for the future. The results are tabulated and returned to the respondents with a second questionnaire based on the results of the first survey. This process may be repeated several times until the respondents collectively generate a consistent view of the future. Only then is the forecast given to the decision makers.

By avoiding face-to-face meetings, neither the respondents nor the interpretation of their views is prejudiced by rank or personality. Might this promote better decision-making in your organisation?

6 Sampling and hypothesis testing

"It is now proved beyond all reasonable doubt that smoking is one of the leading causes of statistics."

Readers Digest

Summary

Sampling saves time and money in a wide range of business situations. Moreover, samples pop up everywhere in unexpected guises. The techniques for handling samples are therefore valuable.

Working with sample data can be highly cost effective. You want to know the average order placed by your customers. You can be 100% sure of the figure by examining all 10,000 invoices, or 99% certain by sampling just 50, which is not a bad trade-off.

Sometimes sampling is inevitable: testing every light bulb to destruction to guarantee a 900-hour life leaves none to sell. The best you can do before buying a machine is to sample its output; you cannot wait until its life cycle is complete and you know with certainty how it performed. This is where it pays to use induction: estimating characteristics of the whole from a sample.

Deduction works in the opposite way. Luigi knows that 15% of his Leaning Tower of Pisa mementos are faulty, so what are the chances that a box of 100 sent to Carlos will contain fewer than 10 non-leaners? Or what are the odds that an as yet unknown sample of executives will overload one of the Star Elevator Company's lifts?

Hypothesis testing is a rigorous way of thinking about sample data to minimise the risks of drawing the wrong conclusions. It throws some light on the decision-making process in general. Note particularly the distinction between a type I and a type II error.

Estimating statistics and parameters

What is a sample?

A sample is part of a parent population. Ten light bulbs from a production line and a handful of wheat seeds from a sack are both samples. They come from populations of many bulbs and seeds. Some things are less obviously samples. An examination of all customer records is really a sampling of all actual and potential consumers.

Significantly, most government figures should be interpreted as samples. For example, total consumers' expenditure figures are compiled from sample information. But suppose a diligent government statistician added the amounts spent by all consumers in a given period. This sum total of actual spending could still be looked at as a sample of all the countless different ways that consumers could have parted with their hard-earned cash. Samples lurk unexpectedly in many time series and cross-sectional data.

How sampling works
Sampling is an extension of descriptive statistics and probability.

1 Sampling handles any summary measures used to describe data: totals, averages, proportions, spreads, and even the shape of the distribution. It allows such characteristics to be estimated: what are the chances that the total weight of any eight passengers will overload this lift? What is the average height of Luigi's towers?
2 Sampling relies heavily on probability's idea of long-run repetition: keep tossing coins and in the long run 50% land heads, so the probability of a head on one toss is 0.5. Keep taking samples, and in the long run the sample will be representative. So take one random sample, and assess the probability that it has the same characteristics as the parent population.

It is not too hard to see that if the average diameter of a flange is 20 inches, the flanges in a large, random sample will probably also average 20 inches. The most likely size to pick is 20 inches, and errors will tend to cancel out.

Confidence in an estimate increases with the sample size. At the logical extreme, there is 100% certainty that a sample including every item in the population has exactly the same characteristics as the parent group. It is common to state an estimate with 99% or 95% confidence, although any level – such as 57.5% – can be used if there is a worthwhile reason.

However, it is generally the absolute sample size that is important, not its magnitude relative to that of the population. A 99% level of confidence is achieved at well below a sample size of 99% of the population. A sample of just 30 items is often adequate.

It is also possible to improve the level of confidence by relaxing the precision of the estimate. It is more likely that a given sample contains flanges averaging 18–22 inches, rather than 19–21 inches in diameter. Come to that, it is pretty certain that the average is 1–1,000 inches. The range is known as the confidence interval, bounded by upper and lower confidence limits.

Thus, there is a fixed relationship between three unknowns: the sample size, the confidence limits and the confidence level. Determine any two, and the third is determined by default.

Randomness

A sample has to be representative if it is to prove its worth. First, the sample and population must be defined carefully. A random sample of *The Times*'s employees would be representative of all its employees, and perhaps the larger population of all newspaper workers. It would not necessarily be representative of all factory workers or of total national employment.

Second, every item in the parent group must have the known random chance of being included in the sample. It would not be sensible to check 24 hours' production by examining just the next ten items from one machine. A politician's mailbag is a biased sample of public opinion, probably containing an inordinate proportion of hate mail, but bias can ensure a representative random selection. To check shop-floor opinion in a firm where only 20% of employees are male, a stratified sample might include exactly 80 women and 20 men chosen randomly from their respective sub-populations. Cluster sampling, incidentally, is a short cut which requires careful design, such as using employees clustered at five companies as a sample of total employment in a particular district.

Randomness is essential, though not always easy to guarantee. One approach is to number items in the population then select from them using computer-generated random numbers, or random number tables. Most PC spreadsheets and programming languages produce random numbers, although occasionally they output the same random sequence on each run.

Sample results concerning people should be viewed with care. People tend to change their minds, which is one reason why surveys of personal opinion, such as voting intentions, tend to go awry. It has been found that opinion polls do not improve in accuracy beyond a sample size of about 10,000.

Calculating standard deviation 2

The box on page 63 outlined the arithmetic that is used when calculating standard deviations. When calculating the standard deviation (or variance) from sample data, you should divide the sum of the squared deviations by the number of observations minus 1, which makes the sample standard deviation (or variance) an unbiased estimator of the population standard deviation (or variance).

For example, for any set of sample data (say 2, 4, 6) the standard deviation (2.0 in this case) is found from the following six steps.

1 Find the arithmetic mean. $(2 + 4 + 6) \div 3 = 4$

2 Find the deviations from the mean:
$2 - 4 = -2$
$4 - 4 = 0$
$6 - 4 = +2$

3 Square the deviations:
$-2^2 = +4$
$0^2 = 0$
$+2^2 = +4$

4 Add the results: $4 + 0 + 4 = 8$

5 Divide by the number of observations minus 1:
$8 \div (3 - 1) = 8 \div 2 = 4$

6 Take the square root of the result: $\sqrt{4} = \underline{2.0}$

Underlying concepts

The concepts underlying sampling are really very simple, although the jargon is unfamiliar. With a little mathematical juggling, it can be shown that the mean and standard deviation of a sample are unbiased estimators of the mean and standard deviation of the population. Take a large sample, say weights of 30 teddy bears. Calculate the mean and standard deviation, say 2.0kg and 0.2kg, and pronounce the mean and standard deviation of the population of bears as 2.0 and 0.2kg. You would not expect these estimates to be spot on every time. It is clear that:

population mean = sample mean ± sampling error.

The error can be quantified. If the sample batch was reselected many times, the mean of each batch would be different. It would mostly be very close to the population mean: sometimes a bit below, sometimes a

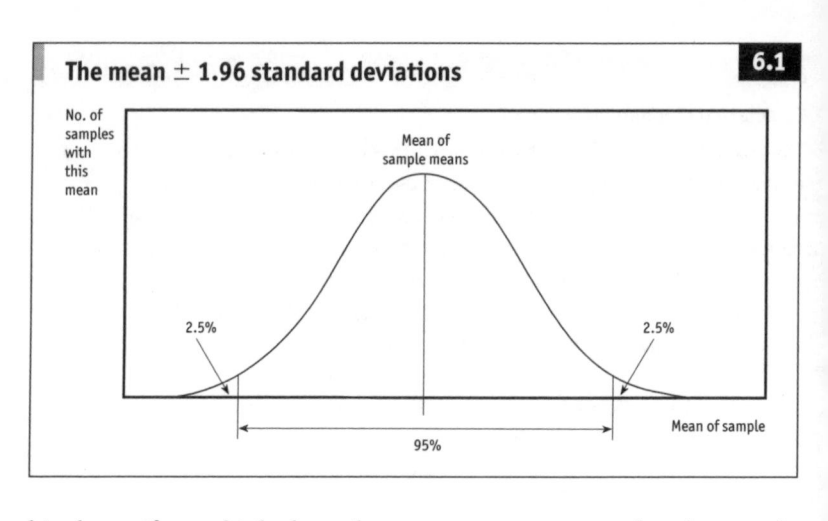

The mean ± 1.96 standard deviations   `6.1`

No. of samples with this mean

Mean of sample means

2.5%   2.5%

Mean of sample

95%

bit above. If you think about the process, you can spot that the sample means would be normally distributed and spread less widely than the total range of values in the population as a whole.

This is the central limit theorem: as the sample size increases, the sampling distribution of the mean of anything approaches the normal distribution, regardless of the shape of the parent population. It is important to grasp the concept of (hypothetical) repeated samplings each with its own mean.

Chapter 3 noted that if you know the mean and standard deviation of a normal population, you know just about all there is to know about it.

The mean of the sample means is taken simply as the mean of the sample. By happy chance, the standard deviation of the sample mean is also easy to calculate. It always equals the standard deviation of the population divided by the square root of the sample size. As it indicates the mean's accuracy as a point estimator, it is usually called the standard error of the mean.

Population standard deviation is frequently unknown, but it can be estimated as equal to the standard deviation of the sample (see Calculating standard deviations boxes on pages 63 and 119). In practice, the population and sample standard deviations are used more or less interchangeably, but wherever possible population data are used to minimise sampling error.

Sample means

About 95% of values in a normal distribution lie within about 2 standard deviations of the mean. A large sample of means is normally distributed and their standard deviation is called the standard error of the mean.

Thus, there is a 95% probability that the mean of a sample lies within 1.96 standard errors of the mean from the population mean (see Figure 6.1).

Confidence

Confidence levels

There is 95% confidence that:

> sample mean = population mean ±
> [1.96 × (population standard deviation ÷ √sample size)]

or, since the same arithmetic works in reverse, there is 95% confidence that:

> population mean = sample mean ±
> [1.96 × (sample standard deviation ÷ √sample size)]

This is exceedingly simple to put into practice. Say you take a random

Sample sums

Estimating a total is a logical extension of estimating a mean. Consider the example in the text where a random sample of 49 wage rates produced a mean wage of $500 with a standard deviation of $28. The 95% confidence interval for the mean wage is $500 ± $7.84. If there are 1,000 employees, the 95% confidence interval for the total wage bill is:

> (1,000 × $500) ± (1,000 × $7.84) = $500,000 ± $7,840.

In other words, there is 95% confidence that the total wage bill is $492,160–507,840. This is quicker and easier to identify than checking the wages for each individual employee, and is accurate enough for many purposes.

Table 6.1 **Useful z scores**

Confidence level (%)	One-way	Two-way
90	z = 1.29	z = 1.64
95	z = 1.64	z = 1.96
99	z = 2.33	z = 2.58

sample of 49 wage rates in a large company, and calculate that the average is $500 with standard deviation of $28. You can say with 95% confidence that the:

population mean wage = sample mean wage ±
[1.96 × (sample standard deviation ÷ √sample size)]
= $500 ± [1.96 × ($28 ÷ √49)] = $500 ± $7.84

From this small sample, you are 95% certain that the average wage in the company is between $492 and $508. (See also Sample sums box.)

Confidence intervals

The confidence interval is not carved in stone. It depends on the choice of z (the number of standard deviations from the mean – see pages 67–9). For a 95% confidence interval, as above, z = 1.96. In general:

sample mean = population mean ±
[z × (population standard deviation ÷ √sample size)] and
population mean = sample mean ±
[z × (sample standard deviation ÷ √sample size)]

Table 6.1 repeats common z scores used in confidence testing (see Table 3.2 for a more complete listing).

Suppose with the wages examples on page 121 you want to state the mean wage to within $5. There are two possibilities, related to the confidence levels and the sample size.

Confidence levels. One option is to ask with what degree of confidence can the initial result be re-expressed with a ± $5 confidence interval.

The basic relationship above rearranges as:

Traps in estimating means

The magic of the normal distribution always works when estimating means from samples of 30 or more items.

For samples of under about 30 items when the population distribution is unknown, other techniques are used (see NON-PARAMETRIC METHODS, page 129ff).

For samples of under about 30 items where the population is known to be normal, z scores can be used only if the population standard deviation is also known. Otherwise, estimating from the variable sample standard deviation rather than the constant population standard deviation results in a flat (platykurtic) distribution.

Fortunately, the mathematician Gosset, who for reasons best known to himself published under the name Student, has done the leg work. His Student's t distribution is used in place of the normal (z) distribution for samples of under 30 from a normal population. Essentially, just treat t's as z's and use Student's t tables in place of normal z tables.

The Student's t distribution starts flat and humps itself towards a normal shape as the sample size increases. Thus, t has a different value for each sample size. However, rather than looking up t against a given sample size n, you have to look it up against n − 1, the divisor used when calculating standard deviation from a sample (see page 119). This n − 1 is called degrees of freedom, or df. For example, a sample of 10 items has 9 df.

One other problem is that when the sample is greater than about 5% of the population size, an error creeps in. As each sample item is selected, the population is reduced in size. The conditional probability of selecting each sample item is more important than with large populations. The variance of the sample is reduced. Adjust for this by multiplying the standard error of the mean by a finite population correction factor, $\sqrt{[(N - n) \div (N - 1)]}$, where N is the population size and n is the sample size. For example, sample mean = population mean $\pm$ [z $\times$ (population standard deviation $\div \sqrt{\text{sample size}}$)] $\times \sqrt{[(N - n) \div (N - 1)]}$.

Many computer programs and statistical references ignore this correction factor on the assumption that the population is nearly always large or infinite. On the other hand, it is safe to include it in all calculations, because it factors itself out as the population size increases relative to the sample size.

$$z = (\text{error term} \times \sqrt{\text{sample size}}) \div \text{standard deviation}$$

The error term is one leg of the confidence interval, \$5 in this case. So

$z = (5 \times \sqrt{49}) \div 28 = 1.25$. Normal tables reveal that where $z = 1.25$, the proportion in the middle of the distribution (ie, between two tails) is almost 80% (see Table 3.2). In fact, you can say with 78% confidence that the mean wage is $500 ± $5. The standard deviation may be calculated using sample data or, preferably, from figures for the population as a whole.
Sample size. The alternative is to ask what sample size must be used to maintain 95% confidence. The relationship rearranges, this time as:

$$\text{sample size} = (z \times \text{standard deviation} \div \text{error term})^2$$

Again, the error term is one leg of the confidence interval: $5. So sample size = $(1.96 \times 28 \div 5)^2 = 121$. A sample of 121 wage rates is required for 95% confidence in estimating the average wage to within ± $5. Again, the standard deviation may be derived from population or sample data.
One- and two-way intervals. The plus-or-minus sign (±) in the basic relationship above indicates a two-way confidence interval. The interval does not have to be symmetrical about the mean. It could just as easily be one-way. With the wages example, you would use the z scores relating to one tail of a distribution if you wanted to be, say, 95% confident that wages are above a given figure (rather than between two amounts).

Other summary measures
The concepts and logic underlying means apply to all other summary measures such as proportions, spread, and so on.

Proportions
Many business situations relate to proportions rather than averages. For example: what percentage of rubber washers have a defective hole? What proportion of workers on the shop floor will vote in favour of strike action?

Proportions are essentially means in disguise. Ten workers were asked if they intended to take industrial action. The results, where 0=no and 1=yes, were 0,1,1,1,0,1,0,1,1,1. The average in favour, 7 out of 10 or 70%, is identical to the proportion 0.7.

A clever bit of maths is available for dealing with 0–1 variables. However, the calculations are so complicated (although not difficult) that it is worth avoiding them wherever possible. Fortunately for relatively large samples (see Pitfalls with proportions box on page 125), the distribution known as the binomial expansion approximates to the normal.

Binomial

The binomial expansion is a little piece of magic for dealing with binomial variables (those with two states such as on/off, 0/1, pass/fail, etc). These occur frequently in business problems.

Consider delivery lorries which are classified as loaded to capacity or partly loaded. If the proprotion π that are full is 0.6 and n = 5 lorries are examined, the probability of x = 3 of those being full is 0.35:

General formula: P(x) $\quad = {}^nC_x \times \pi^x \times (1 - \pi)^{n-x}$

Specific example: P(3 full) $\quad = {}^5C_3 \times 0.6^3 \times 0.4^2$

$\quad = 5! \div (5 - 3)! \div 3! \times 0.216 \times 0.16$

$\quad = 120 \div 2 \div 6 \times 0.216 \times 0.16$

$\quad = 10 \times 0.216 \times 0.16 = 0.35$

The general formula is known as the binomial expansion. The arithmetic can be tedious. Fortunately, when n is greater than 30 and both $n \times \pi$ and $n \times (1 - \pi)$ are greater than 5, the normal distribution can be used in place of the binomial.

Pitfalls with proportions

The problems with proportions are not unlike those with means. The binomial only approximates the normal when n is greater than about 30 and both $n \times \pi$ and $n \times (1 - \pi)$ are greater than 5. Indeed, statisticians worry that if π is not already known approximately, the normal should not be used with samples smaller than 100. If it is, the confidence interval should be calculated with π fixed at 0.5, which produces the most conservative estimate possible.

Also, when dealing with a sample n greater than about 5% of the population, the standard error of the proportion should be multiplied by the finite population correction factor (see Traps in estimating means, page 123).

Where π is the proportion (eg, $\pi = 0.7$):

- mean = sample size $\times \pi$
- standard error of the proportion = $\sqrt{[\pi \times (1 - \pi) \div \text{sample size}]}$
  and, derived from these,
- confidence intervals = $\pi \pm (z \times \text{standard error of the proportion})$
- required sample size = $z^2 \times \pi \times (1 - \pi) \div (\text{error term})^2$
- confidence limit, $z = (\text{error term})^2 \div$ standard error of the proportion

Where the population proportion π is required for determining the sample size or the z score but is unknown, set $\pi = 0.5$. This produces the most conservative estimate.

By way of example, consider that Angel Computers wants to know with 95% confidence what proportion of the business community has heard of its PCs. First, what size sample should it take?

The company has no idea what answer to expect, so it uses $\pi = 0.5$ as its initial estimate. It also decides to estimate the proportion to a $\pm 2.5\%$ (0.025) confidence level. On this basis, the required sample size is $1.96^2 \times 0.5 \times 0.5 \div 0.025^2 = 1,537$.

A random sample of 1,537 consumers reveals that just 27% (0.27) have heard of Angel. So the population proportion may be estimated at $0.27 \pm 1.96 \times \sqrt{[(0.27 \times 0.73) \div 1,537]} = 0.27 \pm 0.02$. Angel has discovered that it can be 95% confident that a mere 25–29% of consumers have heard of its PCs.

Spread and distribution

Means are estimated using knowledge about the way that sample means are distributed around their corresponding population means. Similarly, to make inferences about a standard deviation or a variance, it is necessary to know how the sample variance (s^2) is distributed around the population variance (σ^2).

A common approach is to use the ratio $s^2 \div \sigma^2$, which is known as the modified chi-square (χ^2) after the Greek c, or chi. The ratio has a fixed distribution which changes with the sample size. For small samples, it is heavily skewed to the right. As the sample size increases, the distribution humps towards the normal. The sample size n is usually taken into account by degrees of freedom, $n - 1$ (see Traps in estimating means, page 123). The unmodified chi-square distribution is $\chi^2 = (n - 1) \times s^2 \div \sigma^2$.

Chi-square has its own set of tables (just like z or t tables). However, since the chi-square distribution is unobligingly unsymmetrical, the upper and lower confidence limits have to be determined separately. For example, suppose a random sample of 30 wage rates in a large company suggests that the mean wage is \$500 with standard deviation of \$28. There are $30 - 1 = 29$ degrees of freedom (df). Statistical tables indicate that for 95% confidence at 29 df, chi-square equals 45.7 and 16.0 for the lower and upper limits respectively. The limits themselves are found from:

$$\sqrt{[(\text{degrees of freedom} \times \text{standard deviation}^2) \div \text{chi-square}]}$$

So the lower limit is $\sqrt{[(29 \times 28^2) \div 45.7]} = 22.3$ and the upper limit is $\sqrt{[(29 \times 28^2) \div 16.0]} = 37.7$. Thus, on the sample evidence, there is 95% confidence that the standard deviation of wages in the company is between \$22.3 and \$37.7. The range will narrow if the sample size is increased.

Chi-square is encountered in many other tests. In fact, it is probably one of the distributions most abused by practising statisticians. Chi-square is popular for testing goodness of fit, independence of two variables, and relationships between two or more proportions.

Goodness of fit is interesting. The expected distribution of a sample is compared with the actual results, and chi-square is used to test whether the sample conforms to (ie, fits) expectations. In this way chi-square might be used to determine if total sales are normally distributed or whether regional variations are evenly distributed.

Independence also deserves a mention. Here, chi-square is used to determine relationships, perhaps between the age and sex of shoppers. For example, suppose a retail store wants to find out if there is any link between the age and the sex of customers who purchase particular products. The observed number of customers in each age and sex bracket might be listed in a table and compared with the distribution expected when the two attributes are independent. Chi-square would be used to make the statistical test. This is just a goodness of fit test in disguise. Since the tables are sometimes called contingency tables, independence tests are also known as contingency tests.

It is not possible here to cover the mathematical details of all the tricks with chi-square, but it should be clear that it is a fertile area for exploration.

Two or more samples

Business decision-makers are frequently concerned not with the result of a single sampling, but with the outcome of several. For example, 25% of customers in the north-west react favourably to a new flavour of toothpaste, while the proportion in the south-east is 32%. Is the variation due to sampling error or regional differences in taste? Alternatively, two groups of flanges have certain means and standard deviations; could all flanges have come from the same production batch?

Multiple samplings are also encountered. Suppose several production processes are under review. You need a test to distinguish whether different levels of output are due to sampling variation or to more fundamental performance factors of the machines.

The two basic approaches to these problems are outlined below, although in practice you would probably consult a statistician.

Testing differences. The straightforward approach to comparing two large samples is to use a normal z test. If two samples yield proportions of happy customers of 25% and 32%, you would use a significance test to see if you can be, say, 95% confident that the difference of 7% could not have occurred by chance.

Analysis of variance. For more complex problems, such as those requiring the output of several processes to be compared, analysis of variance (ANOVA) is used. The British statistician Fisher did it first, so the tests are now referred to as F tests. F is just like z, t, and chi-square. The basic procedure is as follows.

- ◪ Calculate the mean for each sample group and then derive the standard error based only on these means.
- ◪ Estimate the variance of the (common) population from which the several sample groups were drawn using only the standard error. This variance is termed MSTR (mean square among treatment groups).
- ◪ Estimate the variance of the (common) population again, this time by pooling the variances calculated independently for each group. This variance is called MSE (mean square error).
- ◪ Check the ratio MSTR ÷ MSE, which forms the F distribution. If the means are not equal, MSTR will be significantly larger than MSE.

There are one or two variations on the design, application and interpretation of F tests. Since these would fill a reasonably large text book, they are not considered here.

Non-parametric methods

The sampling methods discussed so far are by and large parametric methods. They are concerned with estimating parameters using sampling statistics with known distributions, such as the normal distribution; and interval or ratio scale measurements, such as weights and measures. When one of these two conditions is violated, non-parametric methods are required. These are also known as distribution-free or robust methods.

The chi-square test is actually non-parametric since it deals with categorical data.

Where possible, parametric tests are preferred as they are more powerful, just as the mean is preferable to the median. However, non-parametric tests are often wonderful in their simplicity.

For example, the sign test is used to test whether a median is equal to some hypothesised (guessed) value. Simply write down a plus sign (+) for each observed sample value that is larger than the hypothesised median, and a minus sign (–) for each value that is lower. Ignore observations that equal the hypothesised median. The hypothesised median is spot on if the proportion of plus signs is 0.50.

If discussion of ANOVA would fill a volume, any review of non-parametric methods would require a shelf full of books.

Other non-parametrics to be aware of are:

- the Wilcoxon test, which is similar to the signs test;
- the runs test for randomness;
- the Mann-Whitney test of whether medians from two populations are equal; and
- the Kruskal-Wallis test of whether several populations have the same median.

The runs test is interesting, incidentally, not least because it sheds light on risk. Consider tossing a coin ten times. Say it produces the following sequence: H H H T T H H T T H. There are five runs (eg, H H H is the first run, T T is the second, and so on.) The expected number of runs R is approximately (n ÷ 2) + 1, where n is the total number of events. Too many or two few runs suggests that the overall sequence is not random. Most people know that if a fair coin is tossed, there is a 0.5 probability of a head each time, regardless of how the coin landed on the previous toss. A head is no more likely after a run of ten heads than after a run of ten tails. Yet clumping in runs is to be expected, and is predictable to an extent. For example, in about 60 tosses of a coin, a run of five heads and a run of five tails would both be expected. This sort of information can be used to good effect in predicting, say, the performance of a machine.

Hypothesis testing

When using sample data there is a simple procedure which minimises the risk of making the wrong decision from the partial information to hand. The procedure, known as hypothesis testing, adds rigour to decision-making, which is obviously valuable.

The underlying principle

A decision is stated in terms of a hypothesis, or theory. The hypothesis is then tested. It is seldom possible to prove or disprove with 100% certainty the sort of hypotheses which trouble executives, so the risk of incorrect acceptance or rejection is quantified. The decision follows automatically, depending on the level of risk identified relative to that acceptable to the decision-maker.

For example, a bakery will not sell new-formula loaves unless it can reject the theory that: this dough-mix is popular with fewer than six in ten consumers. (Statisticians love double negatives – see below.) The bakers want no more than a 5% risk of making the wrong decision. Suppose they commission a market survey which indicates with 99% confidence that more than 60% of consumers think the new dough is the best thing since sliced bread. There is a 1% risk that the survey was wrong, which is acceptable. So the bakers reject the hypothesis and introduce the new product. There are three points to note.

1 The key hypothesis is always formed cautiously in favour of no change or no action. For this reason it is known awkwardly as the null hypothesis. The onus is always on proving that change has occurred, or that action is necessary. This tends to minimise the risk of rushing off and doing something in error.

2 The risk of falsely accepting the null hypothesis is the opposite of the level of confidence in sample results. This is obvious enough. If there is complete information, there is no risk. If only partial (sample) information is to hand, the inferences drawn from it might be wrong. The risk of being wrong (known as the significance level in hypothesis testing) is the opposite of the probability of being right (called the confidence level in sampling). The boundaries between the two are known as confidence limits in sampling and critical values in hypothesis testing.

3 Hypothesis testing does not require judgments, other than in framing the hypothesis and choosing the acceptable level of risk associated with an incorrect decision. Chapter 7 discusses decision-making techniques which incorporate a greater degree of judgmental input.

The mechanics of hypothesis testing

1 Specify the hypotheses. For example, Evergreen Gnomes determines that it is cost-effective to introduce a new moulding process only if the daily output exceeds 530 models. The null hypothesis is that the average output of the process is less than or equal to 530 gnomes a day (when no action required). Every hypothesis has a mutually exclusive alternative. Evergreen's alternative hypothesis is that the process produces more than 530 gnomes per day (take action to introduce the new process). Rejection of the null hypothesis implies automatic acceptance of the alternative.

2 Determine the acceptable risk. Evergreen might want no more than a 1% risk of accepting the null hypothesis in error.

3 Make the test. This will almost always be a sampling of one kind or another. In this example, Evergreen runs the new process for one month to estimate mean daily output. This is a sample of, say, 30 days' production.

4 Assess the evidence and accept or reject the null hypothesis. Evergreen rejects the null hypothesis and introduces the new system only if the test shows less than 1% risk that the process will churn out 530 or fewer gnomes per day.

Risk, probability and confidence

The three common approaches to hypothesis testing are all based on the same sampling logic, but each highlights a different factor: the confidence interval; the probability of the null hypothesis being true or false; and the critical value of the sample statistic.

For example, if Evergreen ran its new process for 30 days and recorded an output with a mean of 531 and a standard deviation of 3 gnomes, does this imply that the null hypothesis (population mean $\leq$ 530) can be rejected? Consider each approach to hypothesis testing.

Confidence interval. The confidence interval approach is not very useful in practical applications, but it is a good place to start. Evergreen could construct a (100% − risk = 99%) confidence interval for a sample mean of 531 to see if it traps the hypothesised population mean of 530. The lower confidence limit is:

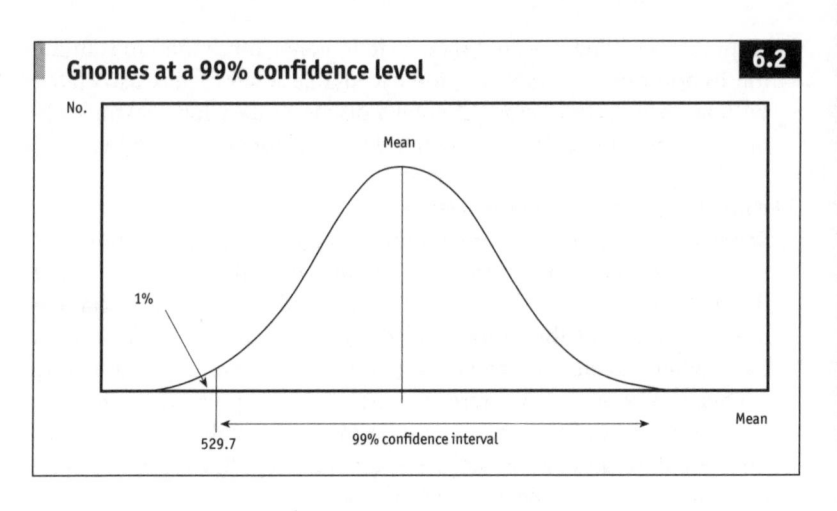

Gnomes at a 99% confidence level 6.2

No.

Mean

1%

529.7   99% confidence interval   Mean

$$\text{sample mean} - [z \times (\text{standard deviation} \div \sqrt{\text{sample size}})]$$
$$\text{or } 531 - [2.33 \times (3 \div \sqrt{30})] = 529.7 \text{ gnomes}$$

The value $z = 2.33$ is found from normal tables, such as Table 6.1. There is, in this instance, no interest in an upper limit. Since the null hypothesis is that the population mean is 530 or less, and this figure is within the 99% confidence interval (529.7 upwards) around the sample mean, the null hypothesis should be accepted (see Figure 6.2).

Probability value. It would be better if Evergreen had a test which highlights the risks surrounding acceptance of the null hypothesis. This risk is revealed, at least in part, by the probability value, or P-value, approach. All Evergreen has to do is find the probability of the sample mean occurring if the null hypothesis (population mean ≤ 530) is true. First, the critical value of z is:

$$\text{(sample mean} - \text{hypothesised population mean)}$$
$$\div \text{(standard deviation} \div \sqrt{\text{sample size}})$$
$$\text{or } (531 - 530) \div (3 \div \sqrt{30}) = 1.83$$

Normal tables, such as Table 3.2 (page 68), reveal that for $z = 1.83$ the one-way confidence level is 96.6%. For a conservative estimate, round this down to 96%. This indicates that if the population mean is indeed 530 or less, there is 96% confidence that an observed mean will be below 531. Conversely, there is a mere 4% significance that the observation will be 531 or above.

Harry's decision options			6.3
	Situation		
	Unsafe	**Safe**	
Decision alternative			
Accept H$_0$,	Correct	Type II	
do not cross	decision	error	
Reject H$_0$,	Type I	Correct	
cross	error	decision	

In other words, there is a 4% probability of observing the sample mean of 531 if the null hypothesis (population mean $\leq$ 530) is true. This 4% is greater than Evergreen's predetermined 1% risk factor. On this basis, the company should accept the null hypothesis and it should not introduce the new process.

Probability value confirms that there is a 4% chance that the null hypothesis is true. The implication, of course, is that there is a 96% risk that it is false.

Probability value is excellent for everyday business use because it highlights the risks around the null hypothesis. So far, however, there is no discussion of the risks of the alternative hypothesis being true or false. Believe it or not, this is not the opposite of the null hypothesis being false or true.

Good and bad decisions

There are two ways to make a right decision, and two ways to make a wrong one. Harry is at the kerb. He must decide whether to cross the road now or later. If he decides to cross, this will either prove to have been a correct decision or he will be run over. Alternatively, if he decides not to cross, this will either be correct, because of a continuous stream of traffic, or incorrect if the road remains empty. The outcomes of Harry's decision process are seen in Figure 6.3.

If the decision to cross is incorrect, Harry makes a fatal error of commission, or type I error as statisticians like to say. If the decision not to cross is incorrect, he makes an error of omission, or type II error. Unless Harry is suicidal, he will prefer to make a few errors of omission (type II) rather than just one error of commission (type I). In other words, Harry attaches a much greater risk to a type I error than a type II error.

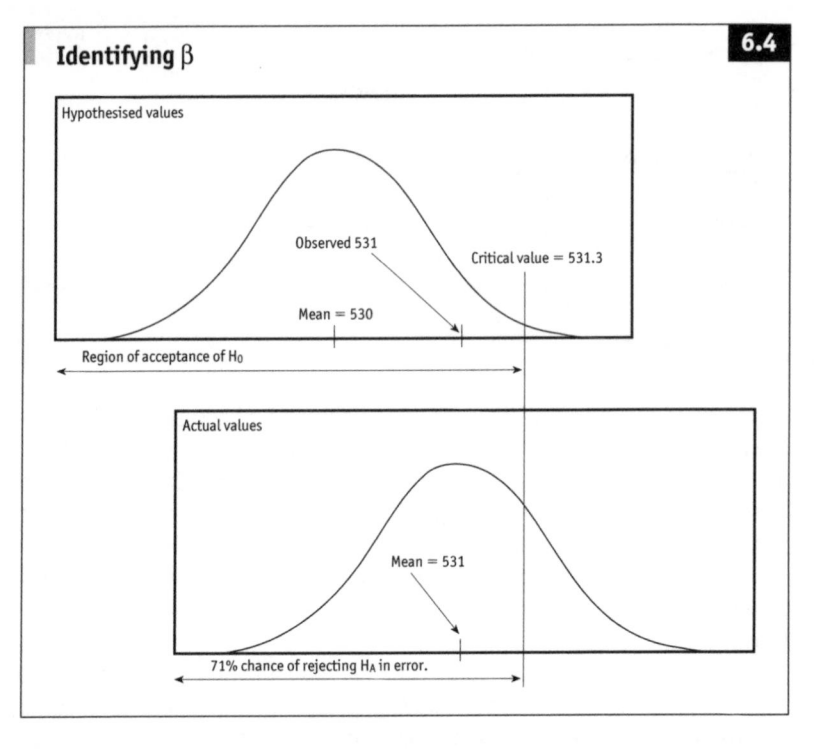

Identifying β

6.4

Hypothesised values

Observed 531

Critical value = 531.3

Mean = 530

Region of acceptance of H₀

Actual values

Mean = 531

71% chance of rejecting Hₐ in error.

However, the choice is not always so clear cut. The trick then is to assess the level of risk attached to each type of error. Logically, the objective is to balance the risks of making type I and type II errors.

Classical decision-making

The risk attached to a type I error of commission always equals the significance level in the test for a null hypothesis. The risk is labelled alpha (α). By convention, alpha tends to be set at $\alpha = 5\%$ or $\alpha = 1\%$. The probability of a type II error of omission is designated beta (β), and is slightly more tricky to quantify. The terminology is easy to remember: type I and alpha both come before type II and beta, while a (or alpha) is the first letter of action and a type I error is the error of commissioning action.

The easiest way to illustrate the application of α and β is to consider the critical value approach to hypothesis testing, better known perhaps as classical decision-making. The essence is to specify carefully the hypothesis and the test.

- ■ Null hypothesis   H₀: population mean ≤ 530
- ■ Alternative hypothesis   Hₐ: population mean > 530
- ■ Level of significance   Alpha, α = 0.01 (1%)
- ■ Test statistic   Sample mean based on a sample of
                                                  30 with standard deviation = 3

This is a one-tail test, where, for 1% in one tail, z = 2.33 (see Table 6.1). The critical value of the sample mean (the boundary between accepting or rejecting the null hypothesis) is:

$$\text{population mean} + [z \times (\text{standard deviation} \div \sqrt{\text{sample size}})] \text{ or}$$
$$530 + [2.33 \times (3 \div \sqrt{30})] = 531.3$$

Since the observed mean of 531 is below the critical value, the null hypothesis must be accepted (see top graph in Figure 6.4). This does not take Evergreen much further than the probability value approach. What is interesting, though, is to look at β, the risk of erroneously rejecting the alternative hypothesis. Suppose the sample estimate was actually spot on, and the population mean is 531. The lower half of Figure 6.4 illustrates this. The risk β is:

$$(\text{critical value of mean} - \text{population mean})$$
$$\div (\text{standard deviation} \div \sqrt{\text{sample size}}) \text{ or}$$
$$(531.3 - 531) \div (3 \div \sqrt{30}) = 0.548$$

Table 3.2 reveals that for a one-tail test where z = 0.548, $\beta \approx 0.71$. Thus, there is a 71% chance of incorrectly rejecting the alternative hypothesis if the population mean is 531 (see Figure 6.4).

The risk of beta error β varies between $1 - \alpha$ (eg, $1 - 0.01 = 0.99$ or 99% for Evergreen) and 0. The worst case (0.99) occurs when the population and sample means coincide. When this happens the high β is generally acceptable since the consequences of a mistake are generally less than when the difference between the sample and population means is great.

The distribution of β for various population means is known as the performance function (or operating characteristics curve) of the test. It is more logical to talk about the power of a test, where power = $1 - \beta$. The higher the power the better. A one-tail test is more powerful, and therefore preferred, to a two-tail test.

There is a snag, however: β and the power of a test can be determined if and only if the population parameter (the mean in this case) is known. Sometimes it is known from previous sampling experiments. When it is not, you might prefer to use the probability value approach.

Conclusion

Sampling theory is perhaps the nub of statistical methods (as opposed to other numerical and mathematical techniques). The economy and sometimes inevitability of working with sample data should by now be clear. The rigour of hypothesis testing helps in drawing accurate and correct inferences from data. More importantly, however, sampling and hypothesis testing contain many pointers that assist with interpreting numbers and making decisions in general.

7 Incorporating judgments into decisions

"A statistician is a man who can go directly from an unwarranted assumption to a preconceived conclusion."

Anon

Summary

Most business decisions involve uncertainty or risk. This chapter outlines techniques which aid decision-making under such conditions and shows you how to bring judgment into the analysis. This allows you to combine the rigour of numerical methods with the strength of managerial experience.

Even when there is complete uncertainty about the business outlook, quantitative techniques provide a framework for thinking about the problems. Decision-making becomes much tighter as soon as the uncertainties can be quantified, no matter how vaguely – when the problem becomes one of decision-making under risk.

This chapter focuses on key techniques for decision-making under uncertainty and decision-making under risk. It looks at how to:

◪ incorporate judgments in the decision-making process;
◪ put in your own evaluation of the desirability of various outcomes;
◪ revise assessments as new information becomes available; and
◪ decide how much it is worth spending to obtain extra information before making a decision.

It concludes with a section on using the normal distribution (introduced on page 65) to tackle complex problems involving multiple decisions and outcomes.

Uncertainty and risk

Identify the problem

First specify exactly what is to be decided. This can almost always be set out in a table with the alternatives down the side and the uncontrollable influences across the top. This brings order to the problem, and often makes the solution seem self-evident.

A family fast food chain, King Burgers, locates a potential new site. Should it open a large or a small restaurant? King's previous experience suggests that a large facility will generate €500,000 profit if the food proves popular, but will incur a €300,000 loss if it falls flat. Similarly, a small takeaway will produce a €275,000 payoff in a good market, but lose €80,000 in a poor market. King's decision matrix is in Table 7.1.

In this example there are two uncontrollable influences (decision analysts like to call these states of nature, or situations), and three decision alternatives (or acts) to worry about. The potential number of rows and columns is unlimited, although other techniques such as decision trees and the normal curve (see below) are better at coping with a large number of alternatives or situations.

Table 7.1 **Basic decision table for King Burgers**

	Uncontrollables	
	Good market €	Poor market €
Decision alternatives		
Large restaurant	500,000	−300,000
Small takeaway	275,000	−80,000
Do nothing	0	0

Table 7.2 **Three decision techniques for uncertainty**

	A	B	C Optimist maximax	D Pessimist minimax	E Average maxiavg	F JEP maxiJEP
Basic decision table						
Restaurant	Good market €	Poor market €	Row maxima €	Row minima €	Row average €	Row JEP €
Large	500,000	−300,000	500,000	−300,000	100,000	−60,000
Small	275,000	−80,000	275,000	−80,000	97,500	26,500
None	0	0	0	0	0	0

Note: The first judgmental expected payoff (JEP), for r = 0.3, is calculated as follows:
JEP = [r × rowmax] + [(1 − r) × rowmin] = (0.3 × 500,000) + (0.7 × −300,000) = −60,000

The body of the decision table is filled with payoffs, that is monetary values allocated to each outcome. Using money wherever possible will sharpen your responses remarkably. If neither cash nor other units (such as sales volume) seem appropriate, allocate scores between 0 and 1 (which are easy to calculate with) or between 1 and 10 or 100 (which might seem more natural).

Note that the do nothing option is a vital ingredient. It has a zero payoff here, because the Kings have built the cost of capital into their profit and loss figures. Often the do nothing payoff is the interest that could be earned by money which would otherwise be spent on the project under consideration.

Decisions under uncertainty

When it is not possible to quantify the uncontrollable influences, there are four simple decision-making techniques available. Suppose that the Kings have no idea what their market is like. For each decision alternative, they consider the outcomes under each state of nature.

1 Optimist King votes for the alternative with the highest possible payoff. He wants to build a large restaurant and hopes for a good market and a €500,000 return. In the jargon, this is maximising the row maxima (maximax). (See Table 7.2, column C.)

2 Pessimist King picks the alternative with the lowest potential loss: do nothing. This conservative approach maximises the row minima (maximin). (See Table 7.2, column D.)

3 Average King assumes that all uncontrollable influences are equally likely, and picks the best return on average: opening a large restaurant (maxiavg). (See Table 7.2, column E.)

4 Hurwicz King, named after a thoughtful mathematician, factors in his own judgment to improve on Average. Hurwicz picks a coefficient of realism (r) between 0 and 1. When he is optimistic about the future, r is near 1. When he is pessimistic, r is near 0. In this case, he judges that there is a mere 30% chance of a good market, so he sets r = 0.3. For each alternative, his judgmental expected payoff (JEP) is:

$$\text{JEP} = [r \times \text{row maximum}] + [(1 - r) \times \text{row minimum}]$$

This produces a weighted average for each decision alternative. The option to plump for is the one with the highest JEP: €26,500 for a small takeaway (Table 7.2, column F).

Table 7.3 **Summary of decisions under uncertainty**

Decision method	Also called	Potential Proposition	payoff €
Optimism	Maximax	Large restaurant	500,000
Pessimism	Maximin	Do nothing	–
Equally likely	Maxiavg	Large restaurant	100,000
Judgmental realism	JEP	Small takeaway	26,500

The recommendations of the four decision-making methods are summarised in Table 7.3. Note that the potential payoffs are not predicted profits; they are merely numbers produced in the search for the best decision. There may be reasons for choosing a decision method – not least the bent of the decision maker. Significantly, though, if the uncertainty can be quantified the Kings move from decision-making under uncertainty to decision-making under risk.

Decisions under risk

Mr King senior has a good feel for the market. His assessment is that there is a 0.6 (60%) probability of a favourable market and a 0.4 (40%) chance of a poor operating environment. The probabilities must sum to 1 since one of these outcomes must occur. This simple assessment of risk opens the way to several analytical techniques.

The Kings can calculate probable long-run payoffs. For each alternative, the expected payoff (EP) is an average of all possible payoffs weighted by the probabilities. For example, the EP from a large restaurant is $(0.6 \times €500,000) + (0.4 \times -€300,000) = €180,000$. Table 7.4

Table 7.4 **Expected payoffs**

	Good market €	Poor market €	Expected payoff €
Probabilities	0.60	0.40	
Large restaurant	500,000	−300,000	180,000
Small takeaway	275,000	−80,000	133,000
Do nothing	–	–	–

Note: The first expected payoff (EP) is calculated as follows:
$EP = (0.60 \times €500,000) + (0.40 \times -€300,000) = €180,000$

indicates that this also happens to be the maximum EP and so is the recommended course of action.

Unlike the potential payoffs thrown up in decision-making under uncertainty, the expected payoff does say something about the Kings' potential profit. The expected payoff is the payoff that would result on average if the decision could be repeated over and over again.

Decision-making always requires an implicit assessment of risks. Once that assessment is explicit the risks are quantified (on a scale of 0 to 1) and the best decision can be identified. Of course, the best decision will vary depending on current, sometimes highly subjective, judgments. When you use this technique, try changing the probabilities to see what effect that has on the expected payoffs. A simple PC spreadsheet decision table makes this what-if modelling easy.

Decision trees

Decision trees are an alternative to decision tables. A tree is a pictorial way of mapping a path through a decision maze. Figure 7.1 is the exact equivalent of Table 7.4. Read from left to right. Rectangles indicate decision points. Circles are chance events. Probabilities are written along the branches stemming from chance events, payoffs are listed at the ends of the branches. Finally, working from right to left this time, expected pay-

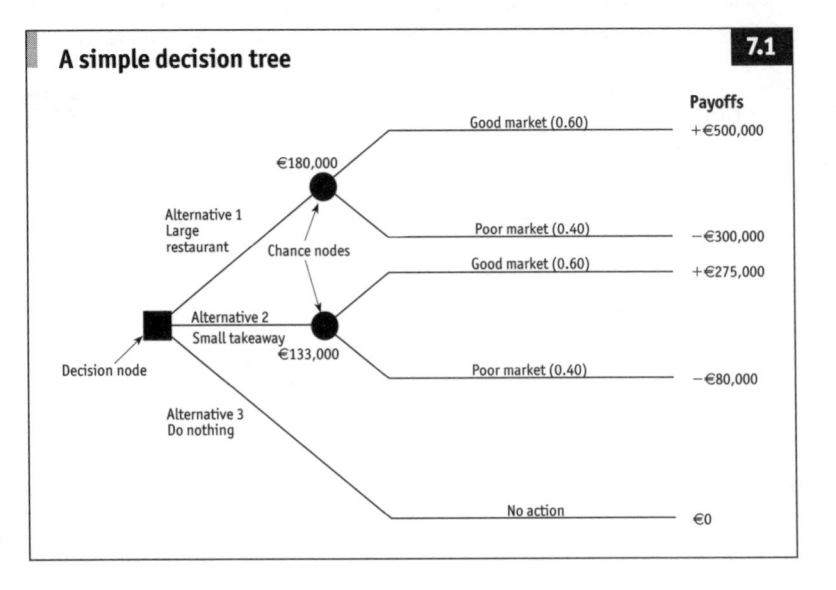

7.1

A simple decision tree

Payoffs

Good market (0.60) — +€500,000

€180,000

Alternative 1
Large restaurant

Chance nodes

Poor market (0.40) — −€300,000

Good market (0.60) — +€275,000

Alternative 2
Small takeaway
€133,000

Decision node

Poor market (0.40) — −€80,000

Alternative 3
Do nothing

No action — €0

offs are noted by each chance event, and the path with the highest expected payoff is easily determined.

The squares and circles shown in Figure 7.1 are known as nodes from the Latin for knot. Perhaps this is why complex decisions are known as knotty problems.

Risk and utility

Expected payoff is an important concept, but there is a snag. Using this approach, the Kings might make a €500,000 profit. But if King senior's judgment is unsound they could instead incur a €300,000 loss. Not surprisingly, this upsets Pessimist King.

To digress for a moment, Frank, who is not an art expert, has a painting. Suppose there is a 0.6 probability that it is the genuine article which he can sell for $2m and a 0.4 chance that it is a worthless fake. A shady, if optimistic, art collector who has not seen the painting offers to buy it on the spot for $1m. Does Frank keep it for the expected payoff of ($2m × 0.6) + ($0 × 0.4) = $1.2m? If so, he might receive nothing. Or does he sell it for a guaranteed $1m? There are times when companies and individuals avoid risk and times when they actively seek it. How can this be factored into the process of decision-making?

For example, what chance of receiving the $2m would you insist on before you would turn down the $1m offer for the painting? Allocate a probability between 0 and 1. If you select a number above 0.5 you are currently a risk avoider. If you chose 0.5 you are risk neutral. If you

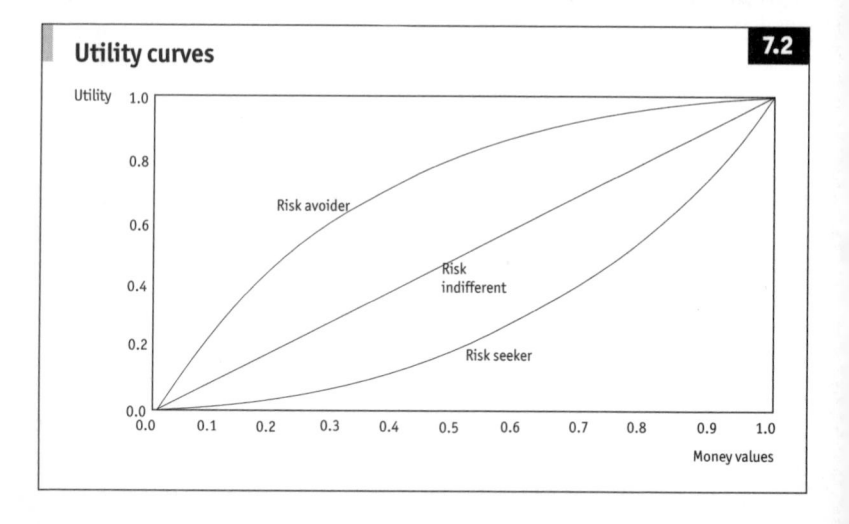

Utility curves   **7.2**

Utility

Risk avoider

Risk indifferent

Risk seeker

Money values

chose below 0.5 you are a risk seeker.

Attitude to risk can be quantified with utility, the economist's concept of the true relative value of various choices. There is no standard measure of utility, but it is convenient to allocate it a score of between 0 and 1, where 1 = maximum utility. The greater utility you attach to a sum of money, the less you are prepared to risk losing it. Utility is subjective; it changes with circumstances, mood, and so on. Financially constrained companies and people generally attach a very high utility to a small sum of money. Three general classes of utility curves are illustrated in Figure 7.2.

The easy way to assess utility is to complete a decision tree of the type in Figure 7.3. This is more colourfully known as a standard gamble. There must be two decision alternatives (ie, keep or sell the picture). The first alternative must have two chance outcomes (genuine painting worth $2m/fake painting which is worthless) and the second must have any certain outcome (receive $1m). For the chance events, the best outcome (a genuine painting worth $2m) is allocated the maximum utility, 1. The worst outcome (a fake, worthless painting) is allocated the minimum utility, 0. You have to select probability P for the best outcome

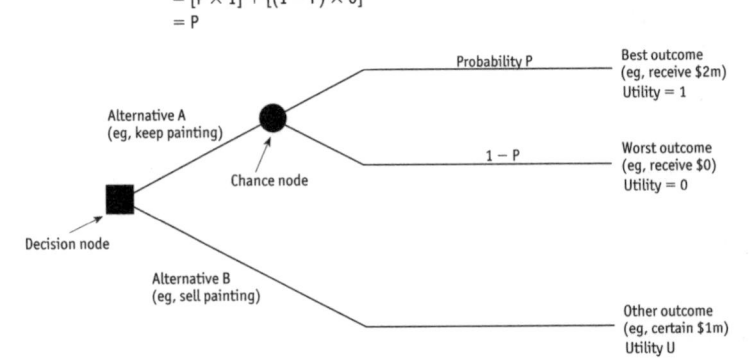

The standard gamble   7.3

Use a utility assessment decision tree with this layout to determine the utilities you attach to any decision alternatives. (Substitute your own alternatives in place of the bracketed items.) Alternative A must have two outcomes with utilities 0 and 1 respectively. Allocate a value to P which makes you indifferent between alternatives A and B. Your current utility for alternative B is exactly equal to P. By simple arithmetic:

utility of alternative A = utility of alternative B
= P × utility of best outcome + (1 − P) × utility of worst outcome
= [P × 1] + [(1 − P) × 0]
= P

Probability P — Best outcome (eg, receive $2m) Utility = 1

Alternative A (eg, keep painting)

Chance node

1 − P — Worst outcome (eg, receive $0) Utility = 0

Decision node

Alternative B (eg, sell painting)

Other outcome (eg, certain $1m) Utility U

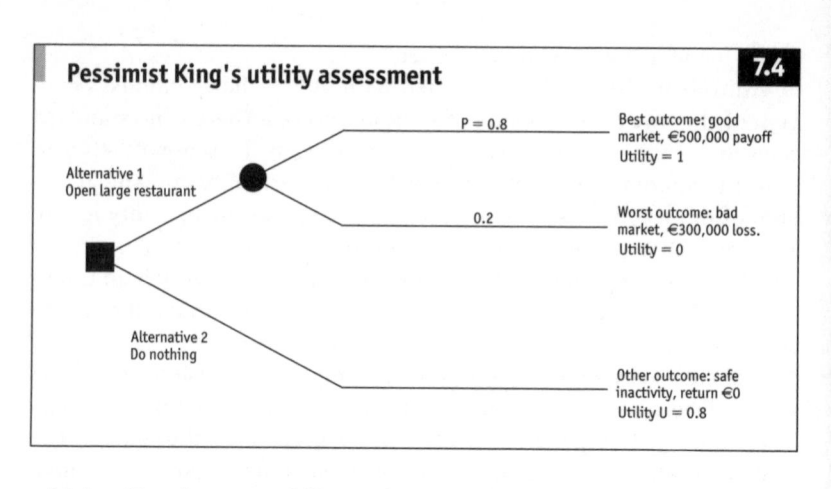

Pessimist King's utility assessment 7.4

Alternative 1
Open large restaurant

P = 0.8 — Best outcome: good market, €500,000 payoff Utility = 1

0.2 — Worst outcome: bad market, €300,000 loss. Utility = 0

Alternative 2
Do nothing

Other outcome: safe inactivity, return €0 Utility U = 0.8

which will make you indifferent between the two alternative decisions. By simple arithmetic, the numerical value of this probability is exactly equal to the utility of the other alternative.

Figure 7.4 shows Pessimist King's utility assessment. Today she requires a 0.8 probability (ie, an 80% chance) of a large restaurant being a success. This gives her a 0.8 utility for the €0 payoff from doing nothing.

The three utilities in Figure 7.4 are plotted on a chart and joined with a smooth curve (Figure 7.5). This is Pessimist King's utility curve. She can read off her utility for each monetary amount in the body of her decision table (Table 7.4). For example, the monetary amount of €275,000 has a utility of about 0.94. The utilities are shown in the first two

Table 7.5 **Expected utilities**

	Good market	Poor market	Expected utility
Probabilities	0.60	0.40	–
Large resturant	1.00	0.00	0.60
Small takeaway	0.94	0.72	0.85
Do nothing	0.80	0.80	0.80

Note: The second expected utility (EU) is calculated as follows:
EU = (0.60 × 0.94) + (0.40 × 0.72) = 0.85
The assessed probability of a good market remains unchanged at 0.60, even though Pessimist King would prefer it to be 0.80.

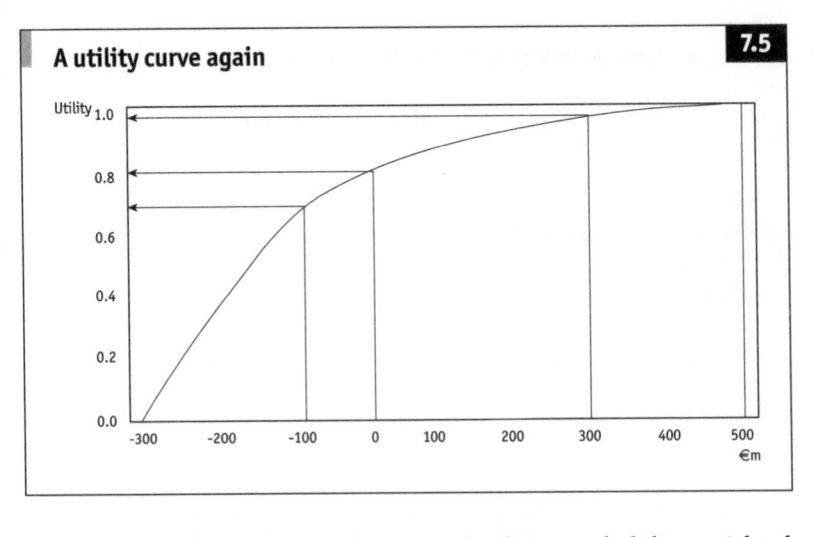

A utility curve again 7.5

columns of Table 7.5 along with expected utilities: probability weighted averages of utility payoffs.

Now, instead of do nothing, the optimal decision (with a utility of 0.85) is to open a small takeaway.

It should be clear that for any decision you can factor in your own utility. Just complete a standard gamble, plot the derived utilities, and read off the figures required for the expected utility table. A PC spreadsheet will do the work, although back of the envelope sketching is usually adequate. Again, it is worth experimenting with different utilities to see what effects they have on your decision process.

The Kings, rather than debating Pessimist's utility function, decide to try to further refine their decision process.

Perfect information

Decisions involving risks depend on an accurate assessment of the risks. In many situations the assessment can be improved through further study. Indeed, the Kings themselves decide to commission some market research. How much should they spend?

If the Kings could have perfect information, they would always make the best decision for any outcome. In a good market they would open a large restaurant. In a bad market they would do nothing. If they could repeat this decision many times, the long-run expected payoff with perfect information (EPPI) would be (0.6 × €500,000) + (0.4 × €0) = €300,000. This is a weighted average of the best outcomes under

Table 7.6 **Expected payoff with perfect information (EPPI)**

	Good market €	Poor market €	Expected payoff €
Probabilities	0.60	0.40	
Large resturant	500,000	−300,000	180,000
Small takeaway	275,000	−80,000	133,000
Do nothing	0	0	0
	↓	↓	
Best outcome	500,000	0 →	300,000 = EPPI

Note: EPPI = (€500,000 × 0.6) + (€0 × 0.4) = €300,000

each state of nature, using prior probabilities as weights (see Table 7.6).

Without perfect information, the Kings were already set for an EP of €180,000. Thus, the additional payoff from having perfect information is €300,000 − €180,000 = €120,000. For any decision, the expected value of perfect information EVPI = EPPI − EP.

Table 7.7 **Revising probabilities**

	Conditional probabilities PC	Prior probabilities PP	Joint probabilities PJ = PC × PP	Revised probabilities PR = PJ ÷ PS
Survey predicts				
good market				
Market actually good	0.75	0.60	0.45	0.85
Market actually poor	0.20	0.40	0.08	0.15
PS (survey results good)			0.53	
Survey predicts				
poor market				
Market actually good	0.25	0.60	0.15	0.32
Market actually poor	0.80	0.40	0.32	0.68
PS (survey results poor)			0.47	

Note: The prior probabilities are King senior's initial assessments.
The first joint probability PJ = 0.75 × 0.60 = 0.45.
The first revised probability PR = 0.45 ÷ 0.53 = 0.85.

Bayes theorem

There is an extremely handy relationship known as Bayes theorem. This is used to revise prior probabilities on the basis of new information. Suppose you know the probability of a series of mutually exclusive events B_1, B_2, B_3 ..., and A is an event with non-zero probability, then for any B_i:

$$P(B_i|A) = \frac{P(A|B_i)\,P(B_i)}{P(A|B_1)P(B_1) + P(A|B_2)P(B_2) + P(A|B_3)P(B_3) + ...}$$

With the King Burger restaurant example, where a positive market is M+, a negative market is M−, and a positive survey is S+, the conditional probability of a positive market, given a positive survey is:

$$P(M+|S+) = \frac{P(S+|M+)P(M+)}{P(S+|M+)P(M+) + P(S+|M-)P(M-)}$$

$$\frac{(0.75) \times (0.60)}{(0.75) \times (0.60) + (0.20) \times (0.40)} = \frac{0.45}{(0.45) + (0.08)} = 0.85$$

Note: $P(A|B)$ is read probability of A given B.

Thus, if the Kings' market survey were to yield perfect information, it would be worth up to €120,000.

The expected value of sample information

Occasionally research yields perfect (or nearly perfect) information. If perfection is expected, EVPI indicates whether commissioning the research is financially worthwhile. A market survey is hardly likely to give perfect information. EVPI can be improved by adjusting it to take into account the likely accuracy of the research.

Managerial judgment and past experience both provide estimates of the accuracy of research. Market research companies frequently boast about their track record. The Kings' market research organisation insists that when it predicts a positive market, it is correct 75% of the time (and wrong 25% of the time). Similarly, when it forecasts a negative market, it is correct in 80% of cases (and wrong in 20%). If the Kings accept these figures (rather than subjectively adjusting them), then the probability of:

Table 7.8 **Summary of King Burgers' revised probabilities**

Probability of	Given	Is
Survey predicting good market		0.53
Survey predicting poor market		0.47
Good market	Prediction good	0.85
Poor market	Prediction good	0.15
Good market	Prediction poor	0.32
Poor market	Prediction poor	0.68

- ◪ predicting a good market and having a good market = 0.75;
- ◪ predicting a good market and having a poor market = 0.20;
- ◪ predicting a poor market and having a good market = 0.25;
- ◪ predicting a poor market and having a poor market = 0.80.

There are two sets of information: these estimates of the likely accuracy of the survey and King senior's own prior probabilities for the operating environment. Combining the two reveals what the survey is likely to show (assuming that his probabilities are correct) and how its accuracy will affect the Kings' decisions.

Table 7.7 shows one way of combining the information. It is also possible to arrive at these results by drawing a tree diagram or using an equation known as Bayes theorem which is useful for automating the process (which is easy with PC spreadsheets).

The new probabilities are summarised in Table 7.8. There are now two decisions: whether or not to commission a survey, followed by the original what size restaurant decision. For such multi-step decisions, the easiest way to proceed is to use a tree. This imposes a logical order on the necessary thought processes.

The Kings' new decision tree is shown in Figure 7.6. Drawing it is straightforward. Starting from the left, the first decision is whether or not to commission the survey. There are three possible outcomes (survey predicts good market/survey predicts poor market/no survey). After any one of these outcomes the (second) decision is, as before: what size restaurant, if any? Note that the second decision is a repeat of the tree in Figure 7.1.

Next, the probabilities just derived are written along the branches, and the payoffs are listed at the end of each branch. Note that the Kings were told that the survey would cost €10,000, so they have reduced each payoff to the right of the commission survey node by this sum.

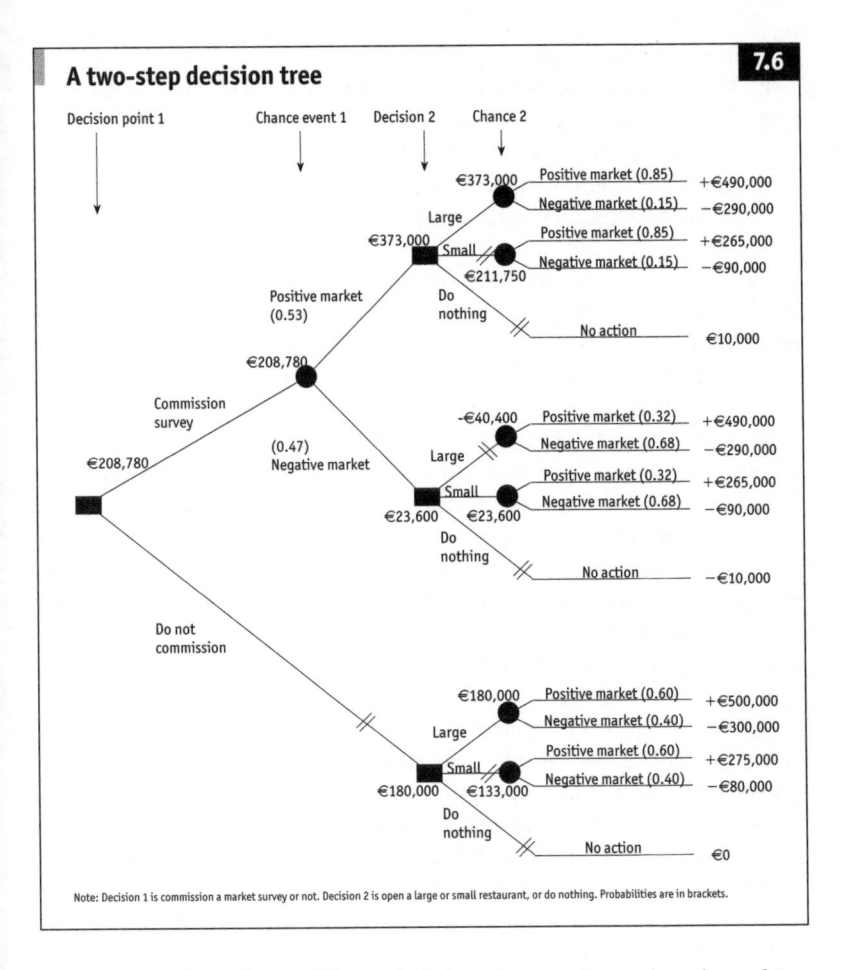

A two-step decision tree · 7.6

Note: Decision 1 is commission a market survey or not. Decision 2 is open a large or small restaurant, or do nothing. Probabilities are in brackets.

Now, working from right to left (starting at the end and working backwards is often a good trick):

- Expected payoffs for the "what size?" decision are calculated as before and entered against each chance node. For example, the expected payoff from opening a large restaurant when the survey result is favourable (top right decision node) is (€490,000 × 0.85) + (−€290,000 × 0.15) = €373,000.
- The highest expected payoffs are taken back to the previous decision nodes. For example, €373,000 is written against the top right decision node since this is the expected payoff from making

the best decision at this point.

◼ The other outcomes are dismissed and their branches are deleted.

◼ The expected payoff is found for the commission survey chance node: (€373,000 × 0.53) + (€23,600 × 0.47) = €208,780.

◼ Again, the highest expected payoff (€208,780) is taken back to the previous decision node, node 1 in this case.

With any tree, by the time that the starting point on the left is reached, the optimum path is self-evident. In this example, the best first decision is to commission a survey (with an expected payoff of €208,780), as opposed to not commissioning a survey.

The optimum path through the tree suggests that a survey would increase the Kings' total expected payoff to €208,780 + €10,000 paid for the survey = €218,780. Without the survey, the expected payoff is €180,000. So the expected value of sample information (EVSI) is €218,780 − €180,000 = €38,780. In other words, it would be worth paying up to €38,780 for the survey. Since it will cost only €10,000, it is worth commissioning. In short, the general relationship is expected value of sample information = expected payoff with sample information less expected payoff: EVSI = EPSI − EP.

Chapter 6 noted that the level of confidence in the results of a sampling exercise (such as a market survey) is directly linked to the size and hence the cost of the sample. This relationship can be exploited when considering whether to commission a survey to improve the inputs to a decision. For example, suppose a decision requires knowledge of the proportion (π) of defective balls produced by a factory. Imagine a tester rushing around a warehouse, bouncing balls to see whether they have the required degree of rebound. The cost (time, etc) of doing this might be 50 cents per ball. The more that are tested, the more accurate the estimate of π, but the greater the cost. The number of balls to test is found by using the relationships illustrated in Figure 7.7.

◼ The cost of sampling (CS) is assumed to increase at a steady rate.

◼ The expected value of perfect information (EVPI) does not change.

◼ The expected value of sample information (EVSI) follows the path illustrated, rising rapidly for the first few samples then levelling off towards its maximum, which is the EVPI.

◼ Thus, the expected net gain from sampling (ENGS = EVSI − CS) is highest at sample size n, at the top of the ENGS hump.

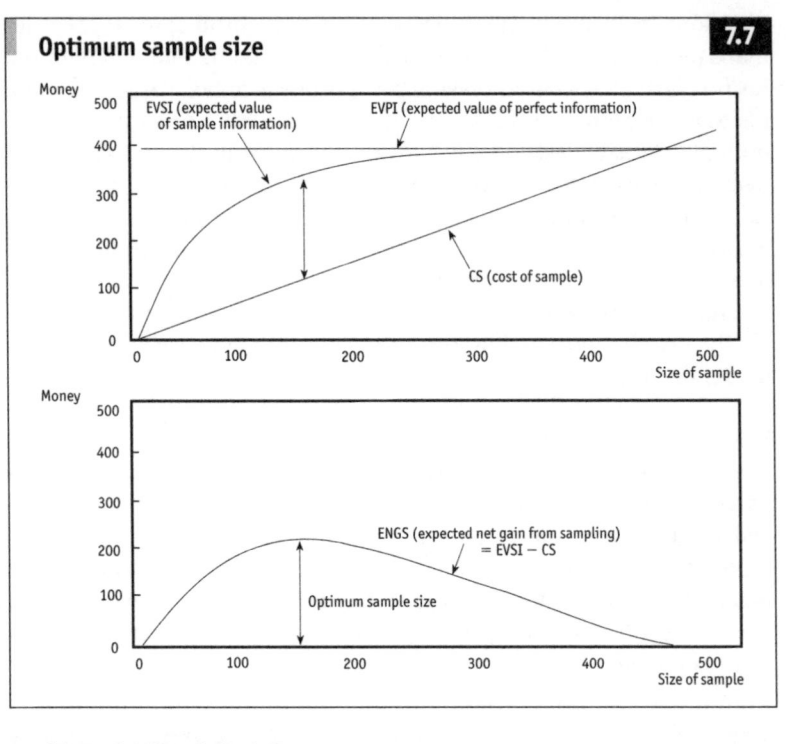

Making the final decision

The Kings started their assessment of risk by estimating likely market conditions. They revised their probabilities on the basis of the expected value of sample information. The sums told them that it would be worth commissioning a market survey costing up to €38,780. The final part of this analysis is to incorporate the survey results, revise the probabilities again, and make the final decision.

There is no need to work through the exercise. Essentially, the survey would provide a new set of probabilities for market conditions. The logic of Table 7.7 or the arithmetic of Bayes theorem would be used to revise the outlook. A decision tree such as Figure 7.6 would be completed and the optimum decision would be read off.

Multiple options

The Kings simplified their decision process by dealing with only three decision alternatives (open a small or large restaurant, or do nothing) and two uncontrollable influences (good or poor market conditions).

Decisions frequently require a rather more thoughtful approach. A sales manager predicting "few" or "many" sales would be told where to put his forecast. The decision-making process needs to be adapted to multiple outcomes (do we invest or not if sales are 0.9m, 1.0m, 1.1m, 1.2m...?), multiple decision acts (do we invest ¥9m, ¥10m, ¥11m, ¥12m ... if demand totals 10m?), or both (how much do we invest for how many sales?). It would, of course, be possible – though not practical – to grow monster decision trees with hundreds or thousands of branches, but there is a much better alternative.

The approach, which is remarkably straightforward, is to use a probability distribution such as the normal distribution (see page 65) to model the uncontrollable outcomes.

Few acts, many outcomes

The following example shows how the normal distribution assists when you have a situation where the number of decision options is limited, but there are a large number of uncontrollable outcomes.

Boosters ("our heel inserts help executives walk tall") has to decide whether or not to introduce a new line in shoe inserts (two decision alternatives). The market might be anywhere between 0 and 1m pairs (many outcomes).

Break even

A business breaks even when its revenues (R) just cover costs (C), where $R = C$.
Revenues = quantity sold multiplied by sale price per unit,

or $R = Q \times P$.

Costs = fixed costs plus quantity sold multiplied by variable costs per unit,

or $C = FC + (Q \times VC)$.

So break even $R = C$ is where:

$$Q \times P = FC + (Q \times VC) \text{ or}$$
$$(Q \times P) - (Q \times VC) = FC \text{ or}$$
$$Q \times (P - VC) = FC \text{ or}$$
$$Q = FC \div (P - VC)$$

Boosters has fixed costs of ¥3.6m, variable costs of ¥600 for each pair of inserts produced, and a selling price of ¥1,000 per pair. Its break even sales level is $3.6m \div (1,000 - 600) = 9,000$ pairs of inserts.

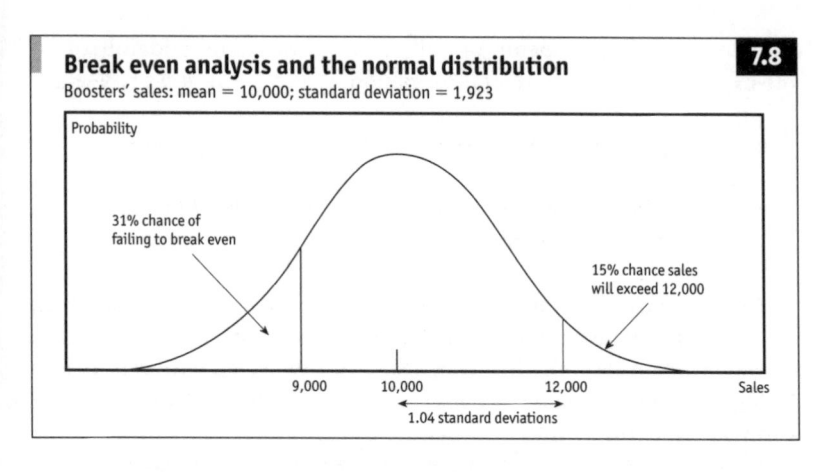

Break even analysis and the normal distribution — 7.8

Boosters' sales: mean = 10,000; standard deviation = 1,923

For this project, the company's fixed costs are ¥3.6m, its variable costs are ¥600 for each pair of inserts produced, and the selling price is ¥1,000 per pair. The marketing manager estimates that Boosters will sell 10,000 pairs next year. She decides that there is a 15% chance that sales will exceed 12,000 and she considers that sales will be distributed normally.

Expected payoff. The first fact is that Boosters' expected payoff (EP) is ¥0.4m: EP = [sales × (price – variable costs per unit)] – fixed costs = [10,000 × (1,000 – 600)] – 3,600,000 = 400,000.

Standard deviation. Next, the marketing manager calculates the standard deviation of her sales forecast. This is done by using z scores and normal tables (Table 3.2). The tables show that z = 1.04 where there is 15% in one tail of the normal distribution (ie, where there is a 15% chance that sales will exceed x = 12,000; see Figure 7.8). Thus, from the key relationships summarised in Table 3.2, 1 standard deviation = (x – mean) ÷ z = (12,000 – 10,000) ÷ 1.04 = 1,923.

Break even. Armed with the mean and the standard deviation, the probability of any level of sales can be deduced. For example, in order to break even (see box), Boosters must sell 9,000 pairs of inserts. First, convert x = 9,000 into a z score: z = (x – mean) ÷ standard deviation = (9,000 – 10,000) ÷ 1,923 = – 0.52. The minus sign simply indicates that z is below (to the left) of the mean. Table 3.2 reveals that for z = 0.52, the proportion in one tail of the normal distribution is 30.15%. Round this to 31% for caution. Thus, there is a 31% chance that Boosters will not break even, and a 69% chance that it will make a profit (see Figure 7.8).

The value of additional information. Finally, the maximum amount that Boosters should spend on additional information is identified by the expected value of perfect information, which is the gap between:

- ◪ the payoff that would result from taking the best possible decision; and
- ◪ the expected payoff from the decision which would be taken without perfect information.

Another way of looking at this gap is to say that it is the loss which results from not taking the best possible decision. This loss is known as the expected opportunity loss (EOL). Since the expected opportunity loss equals the expected value of perfect information (EOL = EVPI), find one and you have found both.

Boosters' EOL is the loss that it will make if it fails to break even. This expected opportunity loss is the total of the expected losses at each level of sales from 0 to 9,000 pairs.

For example, if Boosters sells 8,999 pairs of inserts, it will lose an amount equivalent to the sale price minus the variable costs of production, or ¥1,000 − ¥600 = ¥400. Suppose that the probability of selling exactly 8,999 pairs is about 0.0015 (0.15%). The expected loss at this level of sales is ¥400 × 0.0015 = ¥0.6. Repeat this sum 8,998 more times and add up all the results to find the overall EOL. This is a lengthy, not to say tedious, calculation. Fortunately, since we are dealing with the normal distribution, the work has been done. The final column of Table 3.2 shows the basic results in the form of the unit normal loss (UNL) function. Multiply this by the standard deviation of Boosters' sales to unstandardise it and by the loss per unit to calculate the expected opportunity loss.

Boosters already knows that for sales of 9,000, z = 0.52. Table 3.2 indicates that the UNL for this z is 0.19. The EOL is loss per unit × standard deviation × UNL = 600 × 1,923 × 0.19 = ¥219,222. Or, in other words, the expected value of perfect information is ¥219,222. Boosters should not spend more than this amount on obtaining extra information to assist its marketing decisions. In fact, it should adjust this figure down to take account of the fact that this sort of information is not going to be perfect (as discussed above).

Multiple acts, multiple outcomes

When the number of decision alternatives and outcomes are both large, marginal analysis can help. The next example considers how to determine an optimal inventory level.

Marginal analysis

Most business decisions are made at the margin. With a restaurant selling 305 meals, its strategy may depend critically on how its costs and profits will be affected by the sale of one more cover. In cash terms, it is worth selling the 306th meal if the marginal profit (MP) is greater than or equal to the marginal loss (ML) incurred if the meal is prepared (stocked) but not sold; in shorthand, if MP ≥ ML.

For any sale, the expected marginal profit may be thought of as the probability (P) of making that sale multiplied by the marginal profit: P × MP. Since the chance of failing to make the sale is (1 – P), the expected marginal loss is $(1 - P) \times ML$. So

$$MP \geq ML$$
can be written as $\quad P \times MP \geq (1 - P) \times ML$
or $\qquad\qquad P \times MP \geq (1 \times ML) - (P \times ML)$
or $\quad (P \times MP) + (P \times ML) \geq ML$
or $\qquad\quad P \times (MP + ML) \geq ML$
or $\qquad\qquad\qquad P \geq ML \div (MP + ML)$

In words, this final relationship says that a company should increase its stocks up to the point where the probability of selling one more item is greater than or equal to ML ÷ (MP + ML).

As an example, imagine that it costs the restaurant €10 to prepare a particular dish which sells for €19. For each one sold, the marginal profit is €9. For each one wasted, the marginal loss is €10. From the above relationship, P = 10 ÷ (10 + 9) = 0.53. If the restaurateur's records suggest that there is a 0.55 chance of selling 20 such dishes and a 0.50 probability of selling 21, he should instruct the chef to prepare 20 (ie, where P ≥ 0.53).

Carlos has a tricycle with a cold box. Every day he buys ice creams at €1 each from his wholesaler and cycles down to the beach to sell them for €5 each. By the end of each hot morning any unsold ice creams have to be scooped out and thrown away. He knows that his sales are distributed normally, with a mean of 400 and a standard deviation of 50 ices. (Pages 72ff explains how such estimates are made.) How many should he buy and how many will he sell?

Marginal analysis indicates that it only makes sense to stock an item if the marginal profit (MP) from selling it is at least equal to the marginal

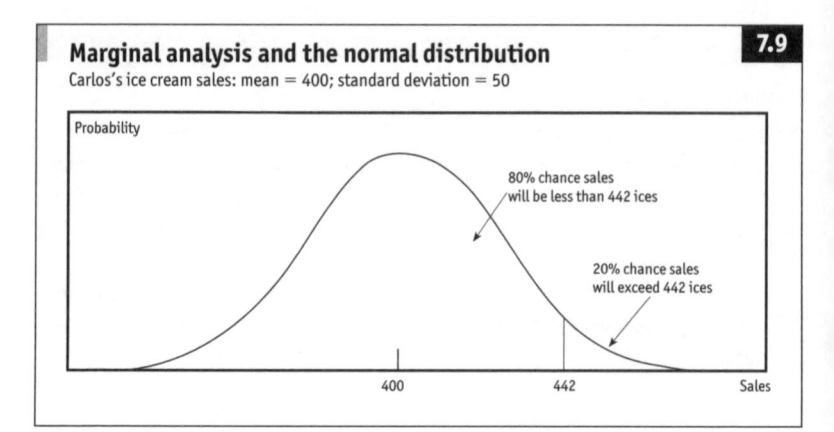

Marginal analysis and the normal distribution 7.9

Carlos's ice cream sales: mean = 400; standard deviation = 50

Probability

80% chance sales will be less than 442 ices

20% chance sales will exceed 442 ices

400   442   Sales

cost (MC) of holding it; or, by the same token, if the expected marginal profit is equal to or greater than the expected marginal cost (marginal loss). This relationship can be developed (see box on page 155) to show that the optimal stock level is where the probability (P) of making a particular level of sales is at least equal to MC ÷ (MP + MC).

For Carlos, the relationship P ≥ MC ÷ (MP + MC) gives P ≥ 100 ÷ (400 + 100) or P ≥ 0.20. In other words, he should not stock ice creams unless there is at least a 0.20 (20%) chance of selling them.

Figure 7.9 shows the distribution of Carlos's sales. The number of sales on the dividing point between the greater than/less than 20% parts of the distribution is found by identifying the z score for 20% in one tail: Table 3.2 shows that at 20%, z = 0.84. The z score relationship x = mean + (z × standard deviation) gives x = 400 + (0.84 × 50) = 442. Carlos should therefore stock up to 442 ices per day.

Conclusion

This chapter has looked at some invaluable decision-making techniques. The final examples using the normal distribution are particularly useful because they model complex real life situations. Do not forget, though, that other distributions – such as the poisson discussed in Chapter 8 –may sometimes be more appropriate.

8 Decision-making in action

"I'll give you a definite maybe."

Samuel Goldwyn

Summary

This chapter explores some specialised (but not specialist) applications of numerical methods. The common thread is the extension of probability and decision techniques. There is not room to do more than scratch the surface here, but this is enough to provide some valuable techniques for everyday use, including the following.

- Game strategy provides a framework for dealing with competitive situations. Among other things, it demonstrates that rational and intelligent action can sometimes result in the worst of all worlds.
- Queueing, which is most readily seen as a rather striking technique for balancing the cost of idle counter staff against the cost of lost customers if waiting lines are too long. It has many unexpected applications beyond dealing with people.
- Stock (or inventory) control has aims similar to queueing but for which there are additional techniques.
- Markov chains help predict what will happen next when that depends on what is happening now.
- Project management is self-explanatory. One useful technique to note is the way that probability distributions can be used to estimate the chances of the project being completed on time and within budget.
- Simulation, which, with a little forethought and a PC spreadsheet, will test business plans without the risks of public failure.

Game strategy

Many people would agree that business is a game. This is significant, since there is a range of numerical techniques that help to improve gaming skills. One, game strategy, contributes to decision-making in competitive business situations.

Two-person matrix game

It is easy to explore this concept with a simple example. Two interior design companies, Corinthian Columns and Regency Row, are moving into a new shopping mall. They can take premises near the centre or on the perimeter of the area. If they both choose central locations, Corinthian estimates that it will win a 60% market share because of its superior reputation. If both companies take end-sites, Corinthian expects to gain 65% of the total business. If Corinthian locates in the middle and Regency at the end, Corinthian hopes to attract 80% of the customers. Lastly, if Regency goes in the middle while Corinthian is on the limits, Corinthian believes that it will get only 40% of the potential business. As always, this information can be summarised in a table (Table 8.1).

In games jargon, this is a two-person matrix game. It is a constant sum game, since the payoffs to Corinthian and Regency always total the same amount (100% of the market share) whatever moves they make; and it is a zero sum game, since one player's gain is the other's loss.

When a player's best option is unaffected by the opponent's move, that player has a pure strategy. When both players have a pure strategy, as in the example here, the game is strictly determined. Industrial espionage will not improve either player's moves.

If you examine Table 8.1, the strategies become clear. Corinthian Columns makes the largest gain (gives up the smallest market share) by locating at the centre of the mall. Regency Row also stands to make the largest gain by opting for a central site.

Their optimal strategies can be reduced to a set of mechanical rules which always work with strictly determined, two person, zero sum games.

◪ Player Columns should identify the maximum value in each column, and should choose the column in which this value is lowest (column 1 – see Table 8.2.)

Table 8.1 **Corinthian v Regency payoff matrix**
% of business Corinthian gives up to Regency

Location	Corinthian Columns	
	Centre	End
Regency Row		
Centre	40	60
End	20	35

Table 8.2 **The game plan**

% of business Corinthian gives up to Regency

Location	Corinthian Columns Centre	End	Row minima
Regency Row			
Centre	40	60	40
End	20	35	20
Column maxima	40	60	

saddle point

- Player Row should find the minimum value in each row, and should choose the row in which this value is largest (row 1).

A strictly determined game has a saddle point: a payoff which is the lowest in its row and largest in its column. This payoff is the value of the game. A game may have more than one saddle point and therefore more than one optimum strategy. In such cases, the payoffs will be identical at each saddle point.

Mixed strategy game

If there is no saddle point, each player's best move will depend on the opponent's move. This is a mixed strategy game. Consider the situation where a game will be repeated many times and neither player knows in advance what move the other will make. To maximise their expected payoffs, each player should make one move in a certain number of games and the other move in the remainder. The optimal strategies can be calculated simply; again, an example will clarify.

Imagine two Russian grocers, Colkov and Rowkov, in an emerging capitalist society. They will sell their cabbages for Rb10 or Rb15 each. Suppose that their payoff matrix is the one shown top-right in the optimal solution for a mixed strategy game (see page 160).

There is no saddle point, no number which is largest in its column and lowest in its row. Since neither can anticipate the opponent's move, they use their imported calculators to determine their optimal strategies using the relationships shown.

For Colkov, $c_1 = 0.27$ and $c_2 = 0.73$. This means that Colkov should aim to sell his cabbages for Rb10 27% of the time, and for Rb15 73% of the time. Similar calculations reveal that Rowkov should

The optimal solution for a mixed strategy game

General rule			*Example*

	Column		**% market share that Rowkov**
	C_1	C_2	**will capture from Colcov**
Row			
R_1	a	b	Price
R_2	c	d	of cabbages Colkov

		Colkov	
		Rb10	Rb15
Rowkov			
Rb10		70	50
Rb15		30	65

**The optimal strategy for player
Column (Colkov) is:**

$C1 = (d - b) \div (a - b - c + d)$ $\qquad (65 - 50) \div (70 - 50 - 30 + 65) = 0.27$

$C2 = (a - c) \div (a - b - c + d)$ $\qquad (70 - 30) \div (70 - 50 - 30 + 65) = 0.73$

Note that $C_1 + C_2 = 1$

**The optimal strategy for player
Row (Rowkov) is:**

$R1 = (d - c) \div (a - b - c + d)$ $\qquad (65 - 30) \div (70 - 50 - 30 + 65) = 0.64$

$R2 = (a - b) \div (a - b - c + d)$ $\qquad (70 - 50) \div (70 - 50 - 30 + 65) = 0.36$

Note that $R_1 + R_2 = 1$

price his vegetables at Rb10 for 36% of the time, and Rb15 for 64% of the time.

This brief introduction to game strategy assumed 2 × 2 games. It is often possible to reduce larger matrices to this size simply by eliminating moves that would never be played because there are rows or columns with better payoffs dominating all positions. Games which cannot be reduced to 2 × 2 format by dominance can be solved by linear programming (see Chapter 9). Games which are not zero sum and those with more than two players require more complex mathematics. It is important to note that game strategy assumes intelligent and rational play. Interestingly, this can result in actions which limit the potential gains from decisions. The prisoners' dilemma illustrates the problem. Briefly, three prisoners in separate cells assess their options.

- If one becomes state witness, he will be freed while his associates are harshly sentenced.
- If they all remain silent, they will receive light sentences for lack of evidence.
- If they all confess, they will receive sterner sentences than if they all kept quiet, but lighter than if only one squealed.

In isolation, they all decide rationally on the first option. As a result, all receive tougher sentences than if they had all kept silent.

Queueing

Next time you are queueing at a supermarket checkout, ponder this. If 95 customers arrive every hour and a single cashier can serve 100 an hour, the average queue will contain 18 irritated people. Opening a second service till would cut the average queue to less than one-third of a person (assuming customers move in turn to the next free cashier). In fact, there are many similar deductions that can be made about queues, all based on straightforward arithmetic and some simplifying assumptions.

Queueing is important. If there are too few service points, queues grow too long and goodwill and custom are lost. If there are too many service points, staff are idle and resources are wasted. Getting the balance just right is critical.

This example considers people in line, but the same logic applies to telephone calls at a switchboard; to trucks arriving to collect or deliver supplies; to trains arriving or aircraft landing; to stocks passing through a production process; and so on.

It is no surprise that queueing has developed its own jargon. Customers in line plus customers in service equals customers in the system. The system may be single channel (one queue behind each service point) or multi-channel (one queue serving many service points). A customer who has to call at more than one service point, queueing more than once, is in a multi-phase system.

Arrivals may be scheduled (eg, one interview every 30 minutes) or random. Scheduled appointments are easy to cope with. Random arrivals are much more interesting. It has been found that they tend to conform to a standard distribution known as the poisson. This is extremely useful. If you know the average arrival rate (10 per hour, 100 per day, or whatever), you can easily estimate other useful information, such as the probability of a given number of arrivals in a particular interval of time, or the probability of a given interval between two

---

Some queueing arithmetic

Using the poisson distribution, if λ is the average number of arrivals, the probability of x arrivals is $P(x) = (\lambda^x \times e^{-\lambda}) \div x!$ where e is the constant 2.7183 and x! is x factorial (see Chapter 1). So if the average number of arrivals is $\lambda = 10$ per hour, the probability of x = 8 arrivals in any given hour is $P(8) = 10^8 \times e^{-10} \div 8! = 0.11$ or 11%.

Similarly, the probability of a delay of between a and b time units between two arrivals is $e^{(-\lambda \times a)} - e^{(-\lambda \times b)}$.

Thus, the probability of a 15–30 minute (0.25–0.50 hour) delay between two arrivals is $e^{(-10 \times 0.25)} - e^{(-10 \times 0.50)} = 0.075$ (ie, 7.5%).

Where the average potential number of servings is μ, the probability that service takes longer than y time units is $e^{(-\mu \times y)}$ (assuming an exponential distribution). For example, if the average service rate $\mu = 3$ per hour, the probability of a service taking longer than half an hour is $e^{(-3 \times 0.50)} = 0.22$ (ie, 22%).

The intermediate results of this arithmetic get rather large, so a spreadsheet is a handy way to tabulate such probabilities.

---

arrivals. The arrival rate is conventionally shortened to a Greek l (lambda) λ.

Service time may be constant (eg, vending machines) or variable (supermarket checkouts). If it is variable it is often assumed to follow another standard distribution called the exponential distribution. Again, one figure (the average number of servings per period of time, say ten an hour) unlocks various estimates, such as the probability of any one service taking longer than a given period of time. The service rate is conveniently shortened to μ.

Clearly, the pivot is the relationship between the number of arrivals λ and the number of people served μ per channel in any unit of time. With a single channel system, the queue is controllable so long as $\lambda \div \mu$ is less than 1 (ie, so long as arrivals are smaller than the number of servings). As $\lambda \div \mu$ approaches 1, the queue grows infinitely long (we have all been in one of these queues). It is useful to recognise that $\lambda \div \mu$ is both the utilisation rate and the probability that the service point is in use. For example, if there are 8 arrivals an hour and capacity is 10 servings, $\lambda \div \mu = 8 \div 10 = 0.8$. Thus the system will run at 80% of capacity on average and there is an 80% chance that the service point is in use at any one time.

Table 8.3 brings together the assumptions about arrivals and service.

Table 8.3 **How long is a queue?**

Number of people likely to be waiting for service

Utilisation rate [a]	No. service points					
	1	2	3	4	5	6
0.10	0.011	0.000	0.000	0.000	0.000	0.000
0.15	0.026	0.001	0.000	0.000	0.000	0.000
0.20	0.050	0.002	0.000	0.000	0.000	0.000
0.25	0.083	0.004	0.000	0.000	0.000	0.000
0.30	0.129	0.007	0.000	0.000	0.000	0.000
0.35	0.188	0.011	0.001	0.000	0.000	0.000
0.40	0.267	0.017	0.001	0.000	0.000	0.000
0.45	0.368	0.024	0.002	0.000	0.000	0.000
0.50	0.500	0.033	0.003	0.000	0.000	0.000
0.55	0.672	0.045	0.004	0.000	0.000	0.000
0.60	0.900	0.059	0.006	0.001	0.000	0.000
0.65	1.207	0.077	0.008	0.001	0.000	0.000
0.70	1.633	0.098	0.011	0.001	0.000	0.000
0.75	2.250	0.123	0.015	0.002	0.000	0.000
0.80	3.200	0.152	0.019	0.002	0.000	0.000
0.85	4.817	0.187	0.024	0.003	0.000	0.000
0.90	8.100	0.229	0.030	0.004	0.001	0.000
0.95	18.050	0.277	0.037	0.005	0.001	0.000
1.0		0.333	0.045	0.007	0.001	0.000
1.2		0.675	0.094	0.016	0.003	0.000
1.4		1.345	0.177	0.032	0.006	0.001
1.6		2.844	0.313	0.060	0.012	0.002
1.8		7.674	0.532	0.105	0.023	0.005
2.0			0.889	0.174	0.040	0.009
2.2			1.491	0.277	0.066	0.016
2.4			2.589	0.431	0.105	0.027
2.6			4.933	0.658	0.161	0.043
2.8			12.273	1.000	0.241	0.066
3.0				1.528	0.354	0.099
3.2				2.386	0.513	0.145
3.4				3.906	0.737	0.209
3.6				7.090	1.055	0.295
3.8				16.937	1.519	0.412
4.0					2.216	0.570
4.2					3.327	0.784
4.4					5.268	1.078
4.5					6.862	1.265
4.6					9.289	1.487
4.8					21.641	2.071
5.0						2.938
5.2						4.301
5.4						6.661
5.6						11.519
5.8						26.373

[a] $\lambda \div \mu$ (ie, average number of arrivals divided by potential number of servings in one period of time). The table assumes multi-channel queueing, poisson arrivals and exponential serving (see text).

It reveals for various utilisation rates ($\lambda \div \mu$) the average number of customers likely to be waiting. In the example just given where the utilisation rate was 0.8, a single channel service system would have a queue of 3.2 people. Two channels would reduce the queue to 0.15 people, and so on. Bank and supermarket managers please take note.

In practical terms, the information helps to determine the best number of service points for a given situation.

There are three key assumptions in this brief outline of queueing theory: poisson arrivals, exponential service time and customers that form only a small proportion of a very large population. The first two can be checked using goodness of fit tests (see page 127). Other assumptions require common sense, such as serving customers in the order of arrival (first in first out, FIFO). Where these assumptions do not hold good, use simulation (see pages 172ff) or, perhaps, finite population queueing tables.

Stock control

The problem of stock (inventory) control is not unlike that of queueing. Too much or too little stock is costly.

The traditional approach assumes that the two main costs are those of holding or carrying stock (C_c) and of ordering (C_o) (see Figure 8.1). For a manufacturing company, the cost of ordering might be thought of as equivalent to the cost of tooling up for a production run to produce the stocks. All other costs, such as the value of the stock itself, are assumed to be constant. Figure 8.1 makes it clear that costs are minimised where order cost $C_o = C_c$ carry cost.

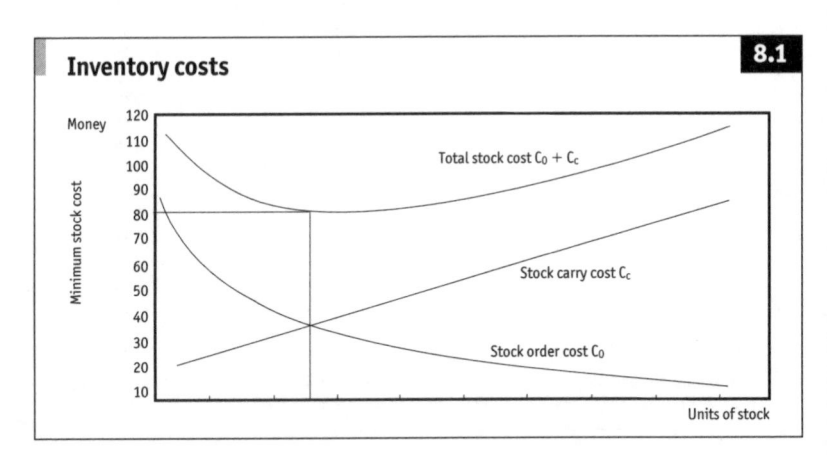

Inventory costs 8.1

Total stock cost $C_o + C_c$

Stock carry cost C_c

Stock order cost C_o

Money — 120, 110, 100, 90, 80, 70, 60, 50, 40, 30, 20, 10

Minimum stock cost

Units of stock

Stock replenishment and consumption   8.2

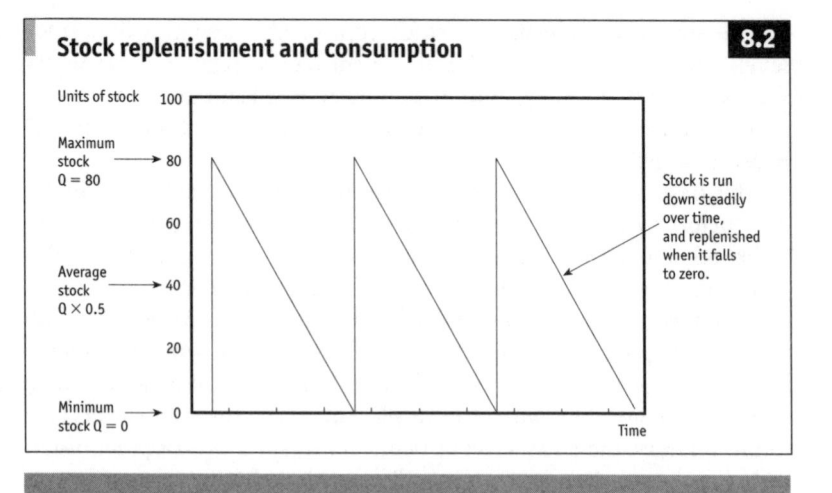

Units of stock   100

Maximum
stock   → 80
Q = 80

60

Average
stock   → 40
Q × 0.5

20

Minimum   → 0
stock Q = 0

Time

Stock is run
down steadily
over time,
and replenished
when it falls
to zero.

Economic order quantity

For any period, such as one year

Where D = annual demand for stock and Q = quantity of stock ordered , the number of orders a year equals D ÷ Q, or D × (1 ÷ Q). The average stock (see Figure 8.2) is Q ÷ 2, or Q × 0.5.

The economic order quantity is where:

Total (annual) ordering costs		Total (annual) cost of carrying stock
or	equals	or
the number of orders times cost per order O_u		average stock times cost per unit C_u:
or		or
$D \times (1 \div Q) \times O_u$		$Q \times 0.5 \times C_u$

The final relationship, $D \times (1 \div Q) \times O_u = Q \times 0.5 \times C_u$, rearranges as:

$$\text{Economic order quantity (EOQ)} = \sqrt{[2 \times D \times O_u \div C_u]}$$

Thus, if a company's annual demand for sprockets is D = 400 units, if the cost of placing one order is O_u = \$100, and if the cost of carrying one sprocket in stock for a year is C_u = \$1.50, the optimal quantity to order at any one time is:

$$\text{EOQ} = \sqrt{[2 \times 400 \times 100 \div 1.50]} = 231 \text{ sprockets}$$

If delivery takes n = 20 days and daily demand is d = 2 sprockets, the reorder point (ROP) is reached when stocks fall to $(n \times d) = (20 \times 2) = 40$ sprockets.

The simple introduction to stock control also assumes that stock levels are run down steadily and replenished at regular intervals by reordering or by running a production batch (see Figure 8.2).

This information is enough to determine the optimal (or economic) order quantity.

This basic model of stock control can be refined considerably. Obvious factors to consider are the cost of holding a buffer stock (safety stock) relative to the costs of shortages (stockout), the value of cash discounts on large orders, and the effects of variable demand.

When demand is not fixed, the trick is to identify its pattern. Chapter 7 noted that sales often follow a normal distribution, while the queueing theory suggests that a poisson distribution may be more appropriate in some situations. By way of example, suppose a restaurateur decides to run no more than a 15% risk of being out of stock of a popular brand of bottled water. Demand is normally distributed with a mean of 400 bottles and a standard deviation of 30 bottles. For 15% in one tail, $z = 1.03$ (see Table 3.2). Thus, there is a 15% chance that demand will rise above mean + (z × standard deviation) = 400 + (1.03 × 30) = 431. So stocks should always be maintained at or above 431 bottles.

An interesting use of marginal analysis and the normal curve for determining optimal stock levels was examined on page 155.

Just-in-time (JIT) stock control aims to avoid an excessive stock build-up. The easy way to visualise this is with a user, a supplier and a storage area requiring just one component. The user starts by taking from the store a container of the parts. When the container is empty, it is returned to storage and exchanged for a waiting second container full of parts. The empty container goes to the supply area for refill and return before the user has finished the second container, and so on. There may, of course, be more than two containers and more than one component. The trick is to balance the flow between storage and producer, and storage and user.

Markov chains: what happens next?

If the stockmarket rose today, what are the chances that it will rise again tomorrow? If a consumer bought EcoClean washing powder this month, what is the probability that he will buy it again next month? If the machine was working this week, what are the chances of a failure next week, or the week after, or the week after that? These are examples of what mathematicians like to call stochastic process: the next state cannot be predicted with certainty. More interesting is the set of stochas-

Table 8.4 **A stockmarket transition matrix**

Probabilities	State tomorrow	
	Rise	Fall
Given state today		
Rise	$\begin{bmatrix} 0.3 & 0.7 \\ 0.6 & 0.4 \end{bmatrix}$	
Fall		

tic processes where the probabilities for the next state depend entirely on the present state. These are known as Markov processes or Markov chains, after the Russian mathematician of the same name.

A good many business problems can be likened to Markov chains. Suppose an investment analyst collects data and concludes that if a share price rises today, there is a 0.7 probability that it will fall tomorrow and a 0.3 chance that it will rise again. Similarly, if it fell today, there is a 0.6 probability of it rising tomorrow and a 0.4 chance of it falling again. The probabilities are summarised in Table 8.4.

This format is known as a transition matrix. The state of the system at any point in time can also be represented by a probability table, or matrix. For example, if share prices rose today, the state matrix for tomorrow is [0.3 0.7] where the two values represent the chances of a rise or fall respectively.

From the transition matrix (call it T) and the state matrix (S), all future states can be predicted. The next state is simply S × T. Table 8.5 shows how S × T is calculated for the stockmarket example. The box expands on the idea of matrices and their multiplication and extends the arithmetic to cover Markov chains with more than two states.

Returning to Table 8.5, the initial state matrix is [1 0] because today's state is known with certainty. The first calculations do not change anything. But the arithmetic for day 2 reveals a state matrix of [0.51 0.49]. This indicates that there is a 51% probability that the share price will rise on day 2, and a 49% chance that it will fall. The probabilities are [0.45 0.55] on day 3, and so on. Notice how by day 5 the state matrix has settled at [0.46 0.54]. In fact, in this case, for any period beyond four days, the state matrix will be the same. The system will have reached equilibrium, a steady state. Equilibrium does not mean static. The state matrix suggests that the share price is likely to decline over time; a pessimistic outlook which, if nothing else, hints at the difficulties of trying to predict equity markets.

Table 8.5 **The stockmarket in transition**

The transition matrix in Table 8.4 permits the following calculations. If the stockmarket rose today (day 0) then the state matrix for tomorrow is as follows.

	State matrix			Transition matrix			State matrix	
Day 1	[1	0]	×	$\begin{bmatrix} 0.3 & 0.7 \\ 0.6 & 0.4 \end{bmatrix}$		=	[0.30	0.70]

From this, the state matrix for the day after tomorrow is

Day 2	[0.30	0.70]	×	$\begin{bmatrix} 0.3 & 0.7 \\ 0.6 & 0.4 \end{bmatrix}$	=	[0.51	0.49]
Day 3	[0.51	0.49]	×	$\begin{bmatrix} 0.3 & 0.7 \\ 0.6 & 0.4 \end{bmatrix}$	=	[0.45	0.55]
Day 4	[0.45	0.55]	×	$\begin{bmatrix} 0.3 & 0.7 \\ 0.6 & 0.4 \end{bmatrix}$	=	[0.47	0.53]
Day 5	[0.47	0.53]	×	$\begin{bmatrix} 0.3 & 0.7 \\ 0.6 & 0.4 \end{bmatrix}$	=	[0.46	0.54]
Day 6	[0.46	0.54]	×	$\begin{bmatrix} 0.3 & 0.7 \\ 0.6 & 0.4 \end{bmatrix}$	=	[0.46	0.54]

For a two-state system such as this, the calculations are:

$$[s_1 \quad s_2] \times \begin{bmatrix} a & b \\ c & d \end{bmatrix} = [(s_1 \times a) + (s_2 \times c) \quad (s_1 \times b) + (s_2 \times d)]$$

Generally, if the two probabilities in the first column of a transition matrix are the same, the system will plunge into equilibrium in period 1. If they are far apart, it will take longer to reach a steady state, perhaps 30 time periods. (Freak transition matrices will never reach equilibrium.) If you can draw up a transition matrix for any problem, you can make neat predictions about the future. One obvious application is translating consumers' inclinations to buy your services or products into potential market shares.

The number of states in the state and transition matrices is not fixed at 2, but since something must happen, every row must sum to 1. PC spreadsheets make it easy to calculate state matrices for a whole sequence of periods. The equilibrium state can then be located by quick inspection and what-if modelling is easy.

Here is a practical use for more advanced Markov analysis: consider a transition matrix for accounts due. States might be paid, due 1 month, due 2 months ... written off. Once a customer pays a bill or the debt is written off, the probability of payment in the future becomes zero.

Matrices

Consider the following pair of bear tables:

Weight (kg)			Bears in stock, by size and colour of fur		
Small	*Large*		*Brown*	*Blue*	*White*
2	3	Small	5	10	12
		Large	6	8	4

It is not too hard to see that the total shelf weight of brown bears in stock is $(2 \times 5) + (3 \times 6) = 28$kg. Similar calculations reveal that the total stock has the following weights:

	Brown	Blue	White
Total weight (kg)	28	44	36

Any table similar to one of these is a matrix. Matrices are usually written without row and column labels just to save time and space. Arithmetic operations are carried out just as if you were combining two tables to create a third. This was an example of matrix multiplication, which clearly is not difficult. It rewrites as:

$$[2 \quad 3] \quad \times \quad \begin{bmatrix} 5 & 10 & 12 \\ 6 & 8 & 4 \end{bmatrix} \quad = \quad [28 \quad 44 \quad 36]$$

Matrices of any size may be multiplied together, but the first (A) must have the same number of columns as the second (B) has rows. Think of the first matrix as being tipped on its side to match the second. Multiply the entries in each row of A by the matching column entries in B, and add the results. Here is one general example with a selection of results indicated:

$$\begin{bmatrix} a & b & c \\ d & e & f \end{bmatrix} \times \begin{bmatrix} o & p & q & r \\ s & t & u & v \\ w & x & y & z \end{bmatrix} = \square$$

$(a \times o) + (b \times s) + (c \times w)$; $(a \times p) + (b \times t) + (c \times x)$

$(a \times r) + (b \times v) + (c \times z)$

$(d \times r) + (e \times v) + (f \times z)$

$(d \times o) + (e \times s) + (f \times w)$; $(d \times q) \times (e \times u) + (f \times y)$

$(d \times p) + (e \times t) + (f \times x)$

Moreover, all accounts will eventually be either paid or written off. This is a situation where many states will eventually be reduced to a few absorbing states (two in this example). Knowing how much money will

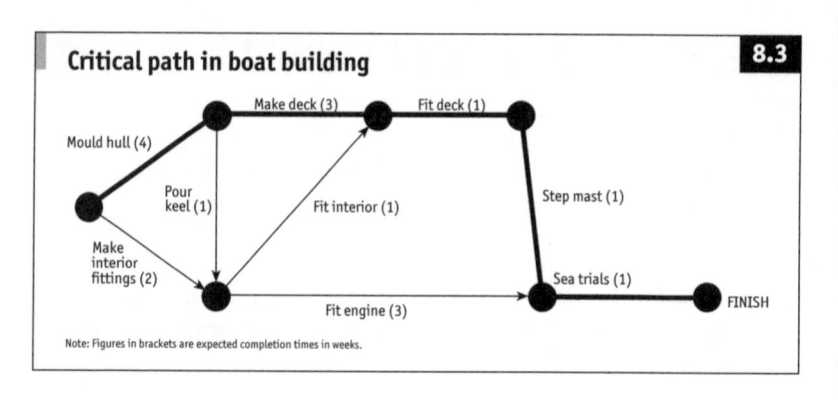

Critical path in boat building 8.3

Make deck (3) Fit deck (1)

Mould hull (4)

Pour keel (1)

Fit interior (1)

Step mast (1)

Make interior fittings (2)

Sea trials (1)

Fit engine (3) FINISH

Note: Figures in brackets are expected completion times in weeks.

end up being paid or in the bad debt category is crucial for managing cash flow.

Project management

To say that accurate project planning is important would be a gross understatement. How many public construction projects can you think of which have overrun their completion date and/or their budgets? Rigorous planning for even quite simple projects can literally pay dividends.

Two well-known (by name, anyway) project planning methods are programme evaluation and review technique (PERT) and critical path analysis (CPA). PERT was developed by the US navy in 1958 and tends to be the method used on that side of the Atlantic. CPA originated at about the same time and is the European equivalent. They are both broadly similar. The main difference is that PERT incorporates probabilities that activities will be completed on time but neglects costs, while CPA works with a normal time (not a normally distributed time) and a crash time. The normal time is the anticipated completion time, and the crash time is the shortest completion time if additional resources are made available.

PERT, CPA and variations all follow similar methods. First, each activity is carefully specified. Orders of priority are identified. ("Lay foundations" before "build walls".) A network is drawn, linking activities and showing priorities. Times and costs are allocated to each activity. The shortest completion time is computed (the critical path) and the network is used for planning, scheduling and monitoring.

Figure 8.3 shows a simplified project for building a sailing yacht. The interior fittings are manufactured while the hull is being moulded and the keel is poured. The interior is fitted while the deck is moulded. When both these activities are complete the deck is fitted, then the mast is stepped.

Activity and project times

The PERT probability approach is useful for reviewing timetables.

For each activity, identify the most likely and the most optimistic and pessimistic completion times. Call these T_M, T_O and T_P. Unless any other model is known to be appropriate, a distribution known as the beta may be used. With the beta distribution, the expected time is a simple weighted average, $[T_O + (4 \times T_M) + T_P] \div 6$. The variance is $[(T_P - T_O) \div 6]^2$. (The 6 reflects the fact that the extremes are ± 3 standard deviations from the mean.)

With the boat in Figure 8.3, the builders say that the most likely moulding time for the hull is 3.8 weeks (where 1 day = 0.2 week), but they also estimate that the most optimistic completion time is 3.2 weeks and the most pessimistic time is 5.8 weeks. Using the simple beta distribution, the expected completion time is $[3.2 + (4 \times 3.8) + 5.8] \div 6 = 4$ weeks and the variance is $[(5.8 - 3.2) \div 6]^2 = 0.2$. This information goes into the overall project planning.

For the overall project, the expected completion time and variance are the sum of these values for the individual activities on the critical path. If activities have expected times of $t_1 + t_2 + t_3 \ldots$ and variances of v_1, v_2, $v_3 \ldots$ the expected completion time for the project is $t_1 + t_2 + t_3 \ldots$ and the overall variance is $v_1 + v_2 + v_3 \ldots$

With the boat-building example, the expected completion time is $4 + 3 + 1 + 1 + 1 = 10$ weeks. If each activity has a variance of 0.2, the overall variance is $0.2 + 0.2 + 0.2 + 0.2 + 0.2 = 1.0$.

The standard deviation for the project is $\sqrt{\text{variance}}$ ($\sqrt{1} = 1$ in the boat-building example).

The individual activities are subject to many small influences, so the normal distribution is good for modelling overall project timing. Since you know the project mean (expected) time and standard deviation, you can derive any information using z scores (see Figure 3.2).

For example, the boat has a mean completion time of ten weeks with a standard deviation of one week. Suppose the new owner requires delivery in 12 weeks. We know that for x = 12:

$z = (x - \text{mean}) \div \text{standard deviation} = (12 - 10) \div 1 = 2$

Table 3.2 shows that where z = 2, there is 2.28% of a normal distribution beyond this point. Thus, on the builders' estimates, there is a modest 2.3% chance that the boat will not be completed in time for the buyer's specified delivery date, and a 97.7% probability that it will be ready.

After the keel is moulded and while other work is continuing, fitters are available to install the engine. Lastly, the yacht undergoes sea trials.

Expected times are noted against each activity. Other numbers are usually added to networks. Such detail is omitted here for clarity. Usually the nodes are numbered for convenience, although this has no particular significance. All activities are tracked. Their earliest start and finish times are noted, beginning from time period zero. In the example, the earliest finish for "make internal fittings" is two weeks, but the earliest start for "fit engine" is five weeks, since it depends also on the completion of the hull and keel.

The longest path through the network is the critical path. In this example, it is ten weeks, which is the shortest possible project completion time.

Following this forward pass, the next step is to make a backward pass. Starting at the end and working backwards, latest start and finish times are noted. For example, the engine must be fitted by $10 - 1 = 9$ weeks. Its latest start time is $9 - 3 = 6$ weeks. So, the engineers can start between weeks 5 and 6, and finish between weeks 8 and 9. With four weeks to complete a three-week job, they have one week of slack time. This may yield resources for activities with no slack.

The forward and backward passes reveal the total project time, the timetable for the various activities, the activities on the critical path which will add to completion time if they are delayed, and the activities which have slack time. Activity and project times show how to estimate the probability that the project will be completed on time.

Budgets for each activity can be combined with the timetable to produce two week-by-week budgets. One is based on earliest start times, the other on latest starts. The earliest start budget will involve spending money earlier, although both budgets will cumulate at completion.

The timetables and budgets are used to monitor and control the project. When it is necessary to crash it (speed it up), the trick is to allocate crash costs per week to the critical path. The activity with the lowest crash cost is the one which is crashed first. But reducing the length of the critical path might make another path critical.

Chapter 9 discusses programming techniques that can be used to find the optimal plan for project crashing. It also looks at some techniques for finding optimal paths through networks.

Simulation

All the numerical methods discussed in this book are focused towards

Table 8.6 **Probability distribution to random numbers**

Cabbages sold per customer	Observed probability	Cumulative probability	Random numbers
0	0.05	0.05	01–05
1	0.10	0.15	06–15
2	0.30	0.45	16–45
3	0.25	0.70	46–70
4	0.15	0.85	71–85
5	0.10	0.95	86–95
6	0.05	1.00	96–100

making better decisions. Only one thing is missing: real life testing. However, the models can even model real life. Just as simulators train and test drivers, pilots and frogmen, so numerical simulation can test business plans.

Imagine a system for controlling stocks. Everything seems fine on paper, but will it work? Could it be that bunching orders will turn tight stock control into a loss of goodwill and of reorder business?

An excellent approach is to generate a sequence of 365 random numbers and use these to represent daily demand over the next year. By comparing this with planned stock controls, you can see whether a stockout might occur, or whether stocks are consistently too high. The exercise could be repeated 100 times to see if the stock policy is always effective. Moreover, certain parameters, such as reorder time, can be changed to see what if. Even rarely encountered events, such as a 1929-type financial crash, can be modelled in.

This sort of simulation is known quaintly as Monte Carlo simulation. This reflects the element of casino-type chance introduced with random numbers. It dates back to the days before computers when mathematicians walked around with pockets full of coloured balls which they used to simulate random events.

The important thing to remember is that real life events are rarely uniformly distributed. If every number between 1 and 100 has an equal chance of popping up in the random sequence, it is not sensible to use these to represent orders for 1–100 units. A customer might be much more likely to buy one cabbage than 20 or 100. Table 8.6 shows how to establish a link between observations (or any standard distribution) and random numbers.

For example, Colkov keeps records of how many cabbages his customers buy. From these records, he calculates the probability of selling a particular number to any customer. If out of every 100 customers, 30 buy two cabbages each, the probability of a randomly selected customer buying two is 0.30 (30%). The probabilities are tabulated in column 2 of Table 8.6. If Colkov did not have historical records, he might guess at the mean and standard deviation of sales per customer and use the normal distribution and z scores to calculate the probabilities (see Table 3.2).

Once the basic probabilities have been found, a running total is calculated (column 3 of Table 8.6) and this is used as the basis for the random numbers as is shown in the table. Note, for example, that if the next random number out of the hat is 14, this simulates the sale of one cabbage.

Next, Colkov might load his PC spreadsheet and put numbers 1–100 in column A. These represent the next 100 customers to enter his shop. He instructs his PC to enter 100 random numbers in column B, and in column C he enters a conditional to convert each random number into values between 0 and 6. For example, it might start IF [the value in cell B1] > 95 THEN [put the value 6 in this cell] ELSE ... and so on.

By this simple process, column C of Colkov's spreadsheet contains 100 numbers in the range 0–6. This is a simulation of his sales to the next 100 customers. Each time he recalculates his spreadsheet, he creates a new set of random numbers and a new simulation of his sales. If he expects to serve ten customers a day and he plans to restock each morning with 20 cabbages, he can see from the spreadsheet the pattern of his overnight stock and the times when he runs out of cabbages before all ten customers have been served. Moreover, he could also use random numbers to simulate the number of customers per day. The possibilities are endless.

Remember, however, that the output of a simulation exercise is only as good as the thought put into setting it up. Also, simulation does not produce optimal solutions in the way of, say, an analysis of economic order levels (see page 165).

9 Linear programming and networking

"If you just torture the data long enough, they will confess."

Anon

Summary
This chapter examines decisions with a more concrete operating background and covers some techniques for handling such decisions. Linear programming is a powerful tool which is used:

◪ when the objective is to maximise or minimise a quantity, such as profits or costs;
◪ when operating under certainty;
◪ when subject to constraints; and
◪ with alternative options.

Linear programming is not necessarily linked to computers. Programming refers to solving a problem numerically, not to computer programs, but computers are good at handling the mechanical work.

Typical programming applications include determining resource allocations to achieve the optimal production mix, portfolio composition, packaging blend, personnel scheduling, policies for advertising, etc.

Linear programming deals with linear (straight line) relationships. Difficulties arise with programming problems which produce curved lines when drawn on a chart, which require integer (whole number) solutions or which have multiple objectives (goals), but there are ways round all these limitations.

The chapter ends with a look at networks and optimisation problems.

Identifying the optimal solution

Linear programming: in pictures

The easiest way to see how linear programming works is with an example. Zak has delivery contracts with two clothing companies. Xperts Inc packs wet weather gear in 0.4m³ cartons which weigh 30kg each when full. The Yashmak Company ships its products in 20kg boxes measuring 0.8m³. Zak's lorry has maximum load capacities of 48m³ and 1,800kg. For historical reasons, he charges Xperts $60 and Yashmak $80 a carton.

How many of each company's boxes should Zak carry to maximise his income? The known facts are summarised in Table 9.1. This suggests three main relationships.

1 Revenues. Zak's income (call this z) will be \$60 for each of Xperts' cartons (call these x) plus \$80 for each of Yashmak's (call these y). In shorthand, ignoring the units which are irrelevant: $z = (60 \times x) + (80 \times y)$.

2 Volumes. 0.4 times the number of Xperts' cartons plus 0.8 times Yashmak's must be less than or equal to 48m³: $(0.4 \times x) + (0.8 \times y) \leq 48$.

3 Weights. 30 times the number of Xperts' cartons plus 20 times Yashmak's must be less than or equal to 1,800kg: $(30 \times x) + (20 \times y) \leq 1,800$. Note also that the number of cartons cannot be negative: $x \geq 0$ and $y \geq 0$.

Bringing all this together to restate the problem in math-speak, the objective is to find values of x and y that make:

$z = (60 \times x) + (80 \times y)$ (known as the objective function) as large as possible, subject to the following constraints:

$$(0.4 \times x) + (0.8 \times y) \leq 48$$
$$(30 \times x) + (20 \times y) \leq 1,800$$
$$x \geq 0$$
$$y \geq 0$$

One way of solving the problem is to plot the constraints on a graph (see Figure 9.1). Note how, for example, all points on the line $(30 \times x) + (20 \times y) = 1,800$ and that all points below the line comply with the constraint $(30 \times x) + (20 \times y) \leq 1,800$. The shaded area in Figure 9.1 contains all the feasible solutions to the problem being considered.

The task remaining is to pick the optimal solution from the set of feasible solutions. Recall that the objective is to make $z = (60 \times x) + (80 \times y)$ as large as possible. This relationship can be rearranged as $y = (z \div 80) - (0.75 \times x)$, which indicates that:

Table 9.1 **Zak's shipping**

	Xperts' cartons	Yashmak's cartons	Zak's van
Zak's income (\$)	60	80	?
Volume (m³)	0.4	0.8	48
Weight (kg)	30	20	1,800

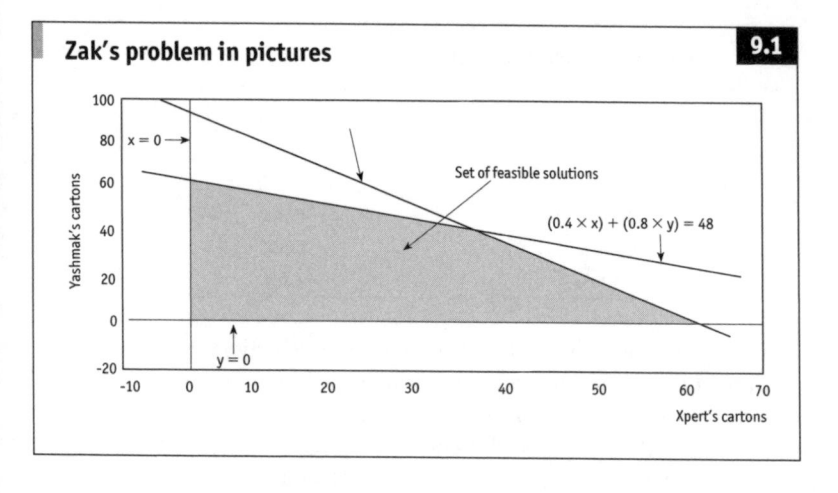

Zak's problem in pictures `9.1`

- for any value of z you can draw a line which has a fixed slope (−0.75); and
- the point at which the line intercepts the x = 0 axis (where y = z ÷ 80) will vary depending on the value of z chosen.

So if you pick some trivial, non-optimal value of z (say 3,200) and plot it on the graph (the lower line in Figure 9.2), you know that the optimal solution will be represented by a line which is parallel to this one. If you roll this trivial solution line towards the top right, z will increase. It is not too hard to see that the optimal solution will have this line as near to the top right as possible, but still touching the set of feasible solutions.

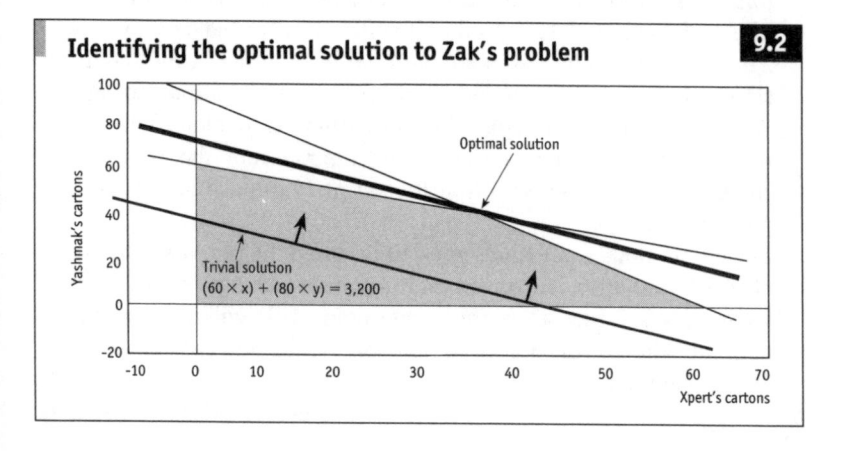

Identifying the optimal solution to Zak's problem `9.2`

Table 9.2 **Corner points from Zak's graph**

Corner points (x , y)	Value of z = (60 × x) + (80 × y) $
(0 , 0)	0
(0 , 60)	4,800
(30 , 45)	5,400
(60 , 0)	3,600

In fact, it is clear that values for x and y which give the largest values of z are at the corner points of the shaded region in Figure 9.2. Table 9.2 lists the corner points (read from the graph) and the corresponding zs (calculated with simple arithmetic). The largest z (5,400) is produced when x = 30 and y = 45. This is the optimal solution.

In the everyday language of the original problem, the largest amount of revenue that Zak can earn on one trip is $5,400. To receive this, he must carry 30 of Xperts' cartons and 45 of Yashmak's.

This has demonstrated a relatively simple way to solve a common type of problem. There are, inevitably, a few points to ponder.

1 When a problem is to minimise rather than maximise, the optimal solution will be a corner point reached when the trivial solution is rolled towards the lower left region of the graph.

2 There may be more than one optimal solution. If two corner points generate the same optimal value of z, then every point on the line joining those corners will also produce that value (see Figure 9.3 A). This is the time to look for other justifications for the decision.

3 z may not have a largest or smallest value, or either. If the set of feasible solutions is unbounded, some corner points are at infinity. In Figure 9.3 B there is no upper bound so there is no identifiable maximal solution. In such a case, you either accept the situation or change/add constraints so that the set of feasible solutions is completely bounded.

4 The corner points rule holds good only when the boundary to the set of feasible solutions is convex (as in Figure 9.3 C). If there is a concave section (as in Figure 9.3 D) the corner points rule will not produce an optimal value for the objective function. Again, you either accept this or change/add constraints.

5 The pictorial approach copes with up to three variables and as many

Tricky linear problems 9.3

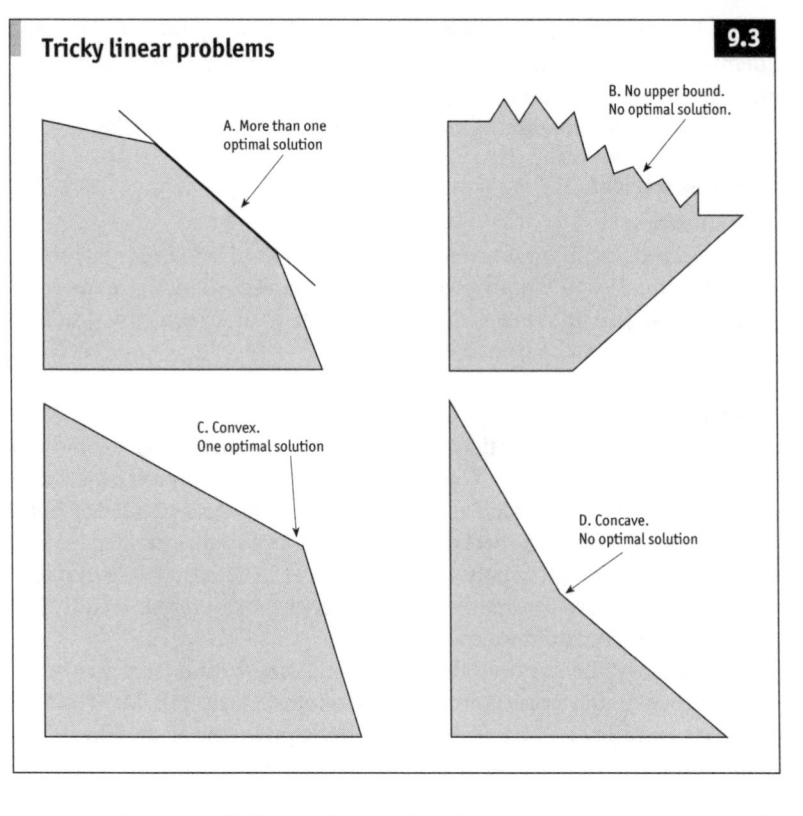

A. More than one optimal solution

B. No upper bound. No optimal solution.

C. Convex. One optimal solution

D. Concave. No optimal solution

constraints as will fit on the graph. Alternative numerical methods are required to deal with more than three variables.

Linear programming: methods and interpretation
The pictorial approach to optimising has a numerical equivalent known as the simplex method. This is particularly useful for dealing with multiple constraints and variables.

The simplex method is a set of rules which can be applied to most linear problems. Many computer packages will do the work, thus making it easy to examine the what-if alternatives of altering the constraints (variables). Some care needs to be taken, however, in interpreting results.

Every linear programming problem can take two different forms. The primal focuses on allocating resources for maximum profit. The dual has a slightly different focus: on the efficient use of resources to minimise costs. The end product is the same, but along the way the dual identifies shadow prices. These reveal, for example, how much of an increase in profits (z) will result from adding one extra unit of resource y. This can be an invaluable aid to management decision-making.

Traps and tricks
Minimisation
The standard linear programming problem seeks to maximise the objective function. When the problem is one of minimisation, there is a very simple trick. Just maximise the z and multiply the result by −1. This sounds too good to be true, but it works.

Integers
With many problems, fractional solutions are not realistic. Zak could not transport half a container. You would not send one-quarter of a salesman on an assignment. Mr Todd cannot make one-third of a teddy bear.

The easiest way to proceed is to find the optimal non-integer solution, and then search for the optimal integer. The result is inevitably approximate. There are methods which give more precise results, but they are complex and time-consuming.

If necessary, be creative. Introduce half-size containers; give your salesman the quarter-assignment plus another task; tell Mr Todd to make an extra bear every three hours, the equivalent of one-third of a bear an hour.

Non-linear relationships
There are some complex methods of dealing with non-linear problems, but a good general approach is to home in on a segment of the problem and use linear approximations to curvilinear relationships. Another possibility is to use transformations.

Multiple objectives
Most programming deals with a single objective. Businesses, however, increasingly have more than one: maximise profits, minimise environmental pollution, maximise market share, and so on. Often these goals are contradictory. If they can be stated in the same units (probably money) linear programming can help. Otherwise, turn to goal programming, a modification of basic linear programming.

The fundamental element of goal programming is that it aims to minimise deviations from set goals. Each goal has two associated deviations, underachievement (d⁻) and overachievement (d⁺). You can also refine the maths by specifying that deviations may be in one direction only (for example, profits may not underachieve the goal) and by ranking the goals in order of priority.

Networks

A challenging class of optimisation problems concerns networks: how to find the shortest route, minimum span, or maximum flow.

Shortest paths

Finding the shortest route from A to B saves time or money. It is also relevant in project management (see pages 170ff). Use Figure 9.4 as an example. How would you find the shortest route if it was not already highlighted? The easiest technique is as follows. Start at the beginning, the origin.

1 Identify the node (knot) nearest to the origin and note its distance from the origin.
2 Look for the next nearest node to the origin which may be a couple of nodes away and note its distance from the start.

Repeat step 2 until you reach the target. The number you noted for this end node is the shortest distance. Backtrack along this final path via the nodes with distances written beside them to identify the shortest route. A more developed approach, incidentally, is dynamic programming; it breaks problems into smaller sub-problems and solves them individually.

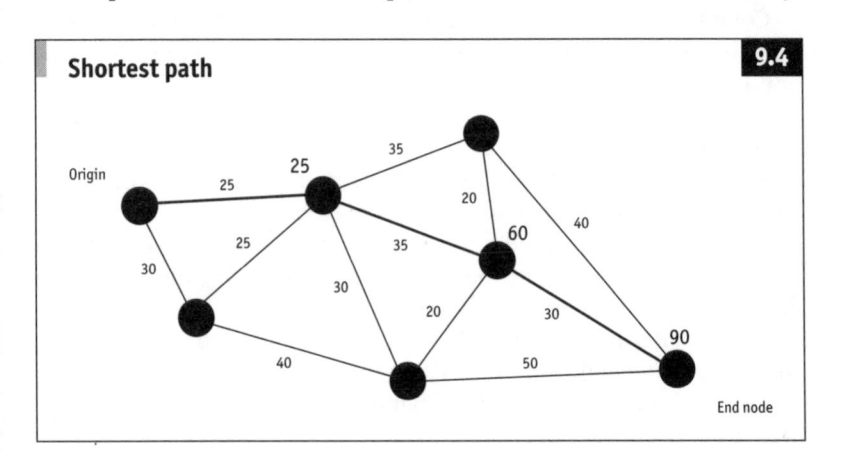

Shortest path 9.4

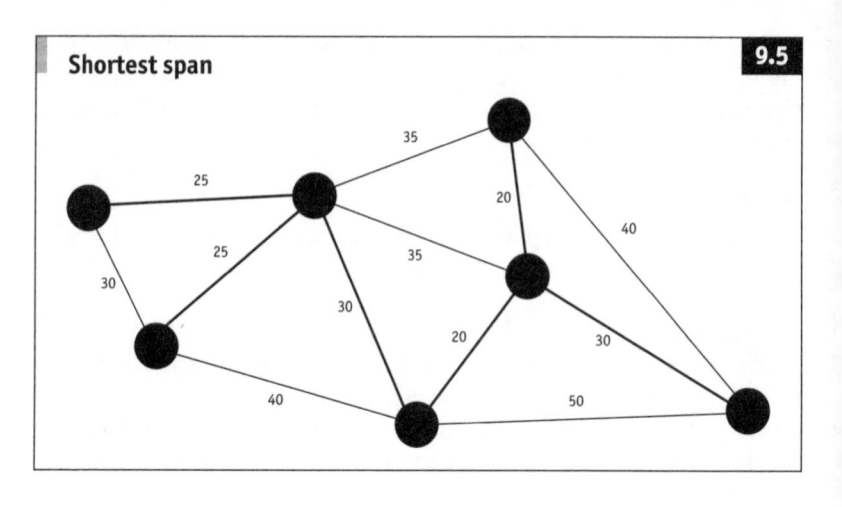

9.5

Shortest span

Shortest spans

Some problems involve distance but not direction. These are sometimes called utility problems, since they are used by utility companies to find the most economical way of connecting consumers to power or telephone networks. The task is to find the shortest span that interconnects a number of points. Figure 9.5 shows an example. Again the technique is simple.

1 Arbitrarily select any node. Connect it to the nearest node.

2 Connect one of this pair of nodes to the unconnected node which is nearest to either of them.

3 Connect one of these three to the nearest unconnected node.

Continue until they are all connected. If there is a tie at any stage, it probably means that there is more than one optimal solution. Just pick one at random and carry on.

Maximum flow

Finding the maximum flow through a system might be used for pumping chemicals or keeping road traffic on the move.

Figure 9.6 shows a road map. Against each node, each path has a number. This represents the maximum flow along that path away from the node towards the next one. In other words, the road from node 1 to node 2 can take up to 300 vehicles an hour travelling from west to east, and 200 moving in the opposite direction.

The maximum flow is found like this. If you want to find the maximum flow from west to east:

Maximal flow 9.6

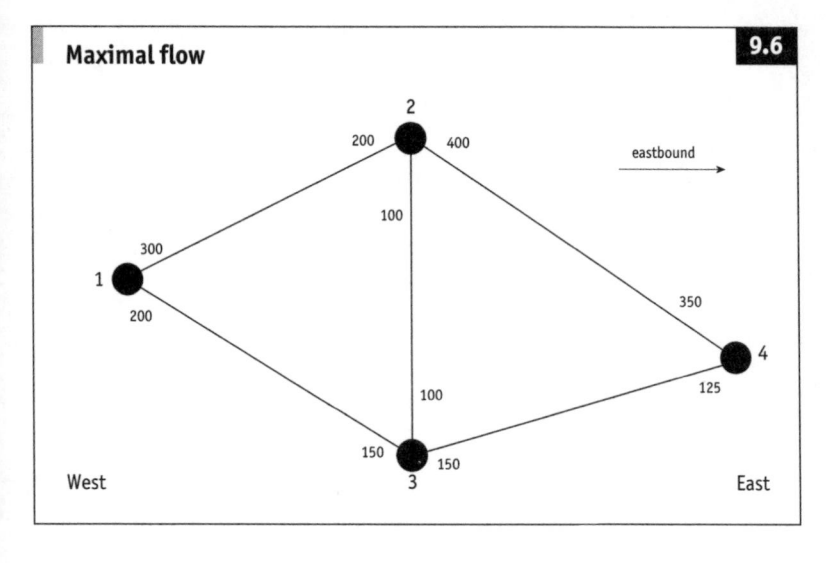

- pick any path;
- identify the highest possible eastbound flow;
- make a note of this amount;
- reduce the eastbound numbers on the path by this amount; and
- increase the westbound ones by the same value.

Keep doing this until there is no path that will take any additional flow. You have reached the optimum solution.

A-Z

Major cross-references are indicated in SMALL CAPITALS. See SYMBOLS for various notations.

A

Above the line. The line is a sum; net income or net profit in income statements or profit and loss accounts. Items that go above the line are normal revenues and expenses of a business. See BELOW THE LINE.

Absolute value. The value of a number ignoring any minus sign. The absolute values of −5 and +5 are both 5. Use absolute values to check the accuracy of, say, a forecast by examining the absolute differences between the projection and the OUTTURN. See MEAN ABSOLUTE DEVIATION and MEAN FORECAST ERROR.

Absorption costing. The inclusion of all FIXED COSTS and VARIABLE COSTS in calculating the cost of producing goods or services. For example, a factory produces one product, say 100,000 cut-throat razors a year. It has fixed running costs of $100,000 a year, and variable costs are $1 per razor for wages, raw materials, and so on. The absorption cost is $2 per item. This might seem to imply that the factory manager should reject the chance to sell an extra 10,000 razors for $1.50 each. But since the variable cost of producing each extra razor is only $1, if the fixed costs are already covered, the order could be profitably accepted. Potential problems are that fixed costs are only fixed over a given range of output and are sometimes difficult to calculate (see IMPUTED COSTS). The opposite of absorption costing is variable (or marginal) costing. See also MARGINAL ANALYSIS.

Accelerated depreciation. DEPRECIATION at a faster rate than that based on an asset's expected life, usually to take advantage of tax concessions.

Accrual. Item relating to one period but paid in another. For example, when costing production in, say, 2003 the accrued fuel costs should be included even if they will not be paid until 2004. See also MATCHING.

Accumulated depreciation. The amount by which the book value of an asset (such as a machine) has been reduced to take account of the fact that it is wearing out or becoming obsolete. See DEPRECIATION.

Acid test. A RATIO used to measure a company's liquidity (also called the quick ratio): cash and debtors divided by current liabilities (UK) or accounts receivable divided by current liabilities (US). A higher ratio suggests greater liquidity, but it may also indicate too much cash tied up in non-productive assets. There is no correct level; compare with other similar businesses. See also CURRENT RATIO.

Adaptive forecasting. Continual tinkering with a forecasting technique to provide a rapid response to changing relationships. Readers of this book will do this automatically by monitoring forecasts against OUT-TURN and asking: "How can I improve the next forecast?"

Algorithm. A set of instructions performed in a logical sequence to solve a problem. See, for example, the SIMPLEX ALGORITHM.

Amortisation. Repayment of a loan (see INVESTMENT ANALYSIS) or DEPRECIATION of a capital asset (such as a machine).

Analysis of variance (ANOVA). A complicated STATISTICAL TEST used to check if differences (variances) exist between the MEANS of several SAMPLES. It might be used to establish whether the AVERAGE output of each of half dozen engines varies due to one factor, perhaps operating temperature (one-way ANOVA); due to several factors, say temperature and fuel additives (two-way ANOVA); and so on; or whether it varies due to chance sampling error. ANOVA does not reveal the size of differences, nor does it confirm that the test is correct; variance might be due to some other unidentified factors.

Annual percentage rate of charge (APR). The UK's statutory EFFECTIVE INTEREST RATE which must be quoted on many credit agreements. APR includes charges such as documentation fees when they are a condition of a loan.

Annuity. A series of regular payments. The arithmetic for dealing with annuities is used for solving most INTEREST RATE PROBLEMS, such as "Should I repay this loan?" or "How much should be transferred to a SINKING FUND?".

Appreciation. An increase in value. The opposite of DEPRECIATION.

Area and volume. In combination, the formulae opposite will solve

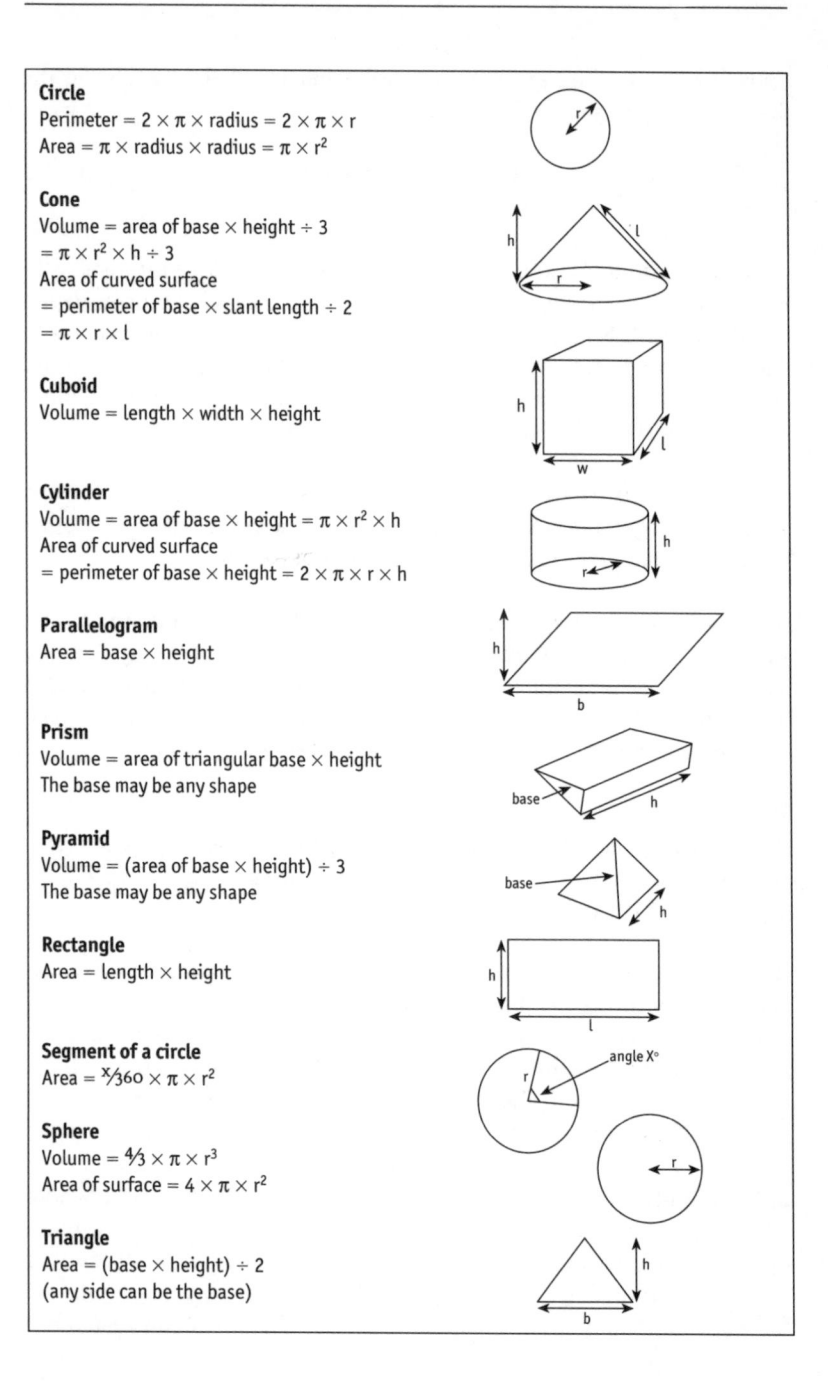

Circle
Perimeter = $2 \times \pi \times$ radius = $2 \times \pi \times r$
Area = $\pi \times$ radius $\times$ radius = $\pi \times r^2$

Cone
Volume = area of base $\times$ height $\div 3$
= $\pi \times r^2 \times h \div 3$
Area of curved surface
= perimeter of base $\times$ slant length $\div 2$
= $\pi \times r \times l$

Cuboid
Volume = length $\times$ width $\times$ height

Cylinder
Volume = area of base $\times$ height = $\pi \times r^2 \times h$
Area of curved surface
= perimeter of base $\times$ height = $2 \times \pi \times r \times h$

Parallelogram
Area = base $\times$ height

Prism
Volume = area of triangular base $\times$ height
The base may be any shape

Pyramid
Volume = (area of base $\times$ height) $\div 3$
The base may be any shape

Rectangle
Area = length $\times$ height

Segment of a circle
Area = $^x\!/_{360} \times \pi \times r^2$

Sphere
Volume = $^4\!/_3 \times \pi \times r^3$
Area of surface = $4 \times \pi \times r^2$

Triangle
Area = (base $\times$ height) $\div 2$
(any side can be the base)

most area or volume problems. Measure heights at right angles to the base. π is the constant 3.1415927 ...

Arithmetic mean. See MEAN. Also GEOMETRIC MEAN, MEDIAN and MODE.

Autocorrelation. The situation when a series of variables is influenced by past values of itself. Departmental budgets are frequently autocorrelated: set at last year's level plus, say, 10%. Analysis of autocorrelation helps to identify patterns in REGRESSION ANALYSIS of TIME SERIES. If there is no autocorrelation remaining, there are no more patterns to tease out. Identify with AUTOREGRESSION ANALYSIS. See also CORRELATION.

Autoregression analysis. Ordinary REGRESSION ANALYSIS used to identify AUTOCORRELATION. Simply regress a series against a lagged version of itself. This is easy with PC SPREADSHEETS.

Average. A descriptive measure which identifies the mid-point of a set of data. The problem is that there are different ways of doing this. The MEAN is in general the best average for everyday use, but see also MEDIAN and MODE. Other measures for special occasions are the GEOMETRIC MEAN, WEIGHTED AVERAGE and MOVING AVERAGE.

B

Bar chart. A pictorial presentation where measurements are represented by the heights of bars. Good for communicating a simple message. See also GRAPHS.

Base. A numerical standard. The base value for an index is the value which is divided into all other values to make the index = 1 (or 100) at the base. For TIME SERIES the base period is frequently a year, perhaps 1995 = 100. Do not be misled by the convergence illusion: the way that two or more indices with the same base will always meet at that base (see Figure 1.2). The base for a set of LOGARITHMS is the number which is raised to a POWER.

Base-weighted index. A WEIGHTED AVERAGE in index form where the weights remain fixed. It is simple to calculate but has two main drawbacks: it overstates changes over time; and it fails to take account of changing relationships. For example, if electronic engineering accounted for 10% of manufacturing output in 1990 and 50% in 2000, a 1990 base-weighted index would fail to reflect the increasing importance of electronic engineering. Sometimes called a Laspeyres index after its founder. Compare with a CURRENT-WEIGHTED INDEX (or Paasche).

Bayes Theorem. A handy formula used to revise PROBABILITIES in the light of new information, described on page 147.

Bear. A person who expects the worst, such as a fall in share prices or exchange rates. The opposite is a BULL.

Below the line. Items below the line are extraordinary items and payments such as dividends. Managers try to boost net income/profit by moving as many items as possible below the line. See ABOVE THE LINE.

Bernoulli variables. VARIABLES with two states (on/off, good/bad) also known as binary, BINOMIAL, switching, 0/1, etc. Many business problems break down into a choice between two options which can be represented by Bernoulli variables. Special techniques are available for handling such variables (see, for example, page 125).

Beta distribution. A standard DISTRIBUTION useful for MODELLING the COMPLETION TIMES of individual activities. See ACTIVITY AND PROJECT TIMES, page 171.

Beta risk. Risk of an ERROR OF OMISSION. The beta coefficient of a share shows its sensitivity to changes in the prices of shares in general.

Binomial. The binomial expansion is a practical way of dealing with binomial variables (those with two states such as on/off, 0/1, good/bad) which occur frequently in business problems. The calculations can be tedious. Fortunately, when certain conditions are fulfilled the much more simple NORMAL DISTRIBUTION can be used in place of the binomial (see page 125).

Break even analysis. Identification of the sales volume required to cover variable cost per unit plus fixed overheads. In combination with expected sales modelled by, say, the NORMAL DISTRIBUTION, break even reveals the PROBABILITY of reaching any level of profit or loss.

Budget. A plan of revenue and expenditure by department, cost centre or profit centre, usually month-by-month for one year ahead. Differences between budget and OUTTURN are the accountant's VARIANCES, which identify deviations from plan and highlight any need for corrective action or revision of the budget for the remainder of the year. Identify how much of any variance is due to INFLATION and how much reflects changes in volumes (see VALUE AND VOLUME). Iron out distortions due to SEASONALITY by comparing each month with the same month of the previous year. Be careful that rigid adherence to a budget does not override long-term common sense. A budget is a plan, not a forecast, but FORECASTING techniques help with preparation of the plan. SPREADSHEETS can also assist.

Bull. A person who expects a favourable event, such as a rise in sales or share prices. The opposite is a BEAR.

C
Capital. A sum of money (PRINCIPAL) placed on deposit; the cash used to buy an asset (see INVESTMENT ANALYSIS); or a measure of a company's finance (total assets or total assets less short-term liabilities or shareholders' equity, etc: see COST OF CAPITAL).

Cash flow. In general, the net flow of money into and out of a business during a particular period. In the USA it is net income plus depreciation, which gives a broad indication of cash earned by trading activities.

Catastrophe theory. A QUALITATIVE planning technique which helps to prepare for sudden, large changes (not necessarily disasters) which cannot be predicted by standard numerical FORECASTING techniques. All businesses should be prepared for a sudden jump (up or down) in, say, sales perhaps due to technological change or civil unrest.

Categorical data. Information in categories, such as male/female. Compare with ORDINAL DATA (eg, poor/fair/good) and INTERVAL DATA (eg, heights and weights).

Causal forecasting. Projecting the future on the basis of established relationships: cause and effect. Sales might be found to depend on incomes, tax rates and population, perhaps with a LAG. A decrease in personal taxation in January might show up in higher sales in the following June. Use common sense and REGRESSION ANALYSIS to identify relationships, but remember that what happened last year might not hold good this year.

Central limit theorem. Stat-speak for the way that any DISTRIBUTION of sample MEANS becomes normal as the SAMPLE size increases. It means that the normal distribution can be used for estimating a mean when the sample size exceeds about 30 items (otherwise use the t-distribution: see t-TEST).

Chi-squared test. A STATISTICAL TEST for establishing CONFIDENCE that the VARIANCE of a SAMPLE is representative of that of the parent POPULATION. Popular for GOODNESS OF FIT TESTS and INDEPENDENCE TESTS.

Cluster analysis. A method of coping with a large number of VARIABLES. Data are grouped into clusters to reduce the number of items in an EQUATION. For example, if you have details of children's spending habits collected from a large number of schools, you might aggregate data from similar schools to reduce the amount of information to a manageable level. The trick is making sure that the clusters are made up sensibly. See also MULTIVARIATE ANALYSIS.

Cluster sampling. A SAMPLING short-cut which requires careful handling to avoid biased or distorted results. Employees clustered at a handful of firms might be sampled to provide evidence about all employees in a particular district. The clusters might be identified using MULTISTAGE SAMPLING techniques. Other sampling methods include QUOTA and STRATIFIED SAMPLING. See INFERENCE for a more general discussion of decision-making using sampling.

Coefficient. Fancy name for a CONSTANT. In an EQUATION such as profit = (20 × units sold) − costs, where 20 might indicate a sale price of $20 per item, 20 is the coefficient of sales. With $(2 \times x) + (10 \times y) = 30$, the coefficients of x and y are 2 and 10. The term is also applied to some descriptive statistics, as follows.

- ◪ Coefficient of correlation. Better known as r. Measure of CORRELATION between two series.
- ◪ Coefficient of determination. Better known as r^2. Byproduct of REGRESSION ANALYSIS which indicates how much of the

DEPENDENT VARIABLE is explained by the INDEPENDENT
VARIABLE. See CORRELATION.

◼ Coefficient of serial correlation. See DURBIN-WATSON.
◼ Coefficient of skew. A measure of SKEW in a distribution.
◼ Coefficient of variation. A standardised STANDARD DEVIATION
  for comparing two standard deviations. Divide the standard
  deviation by the MEAN and multiply by 100 (this gives the
  standard deviation as a PERCENTAGE of the mean).

Collinearity. A high degree of CORRELATION between two VARI-
ABLES. If in REGRESSION ANALYSIS two INDEPENDENT VARIABLES
exhibit collinearity (ie, are highly correlated) it is difficult to tell what
effect each has on the DEPENDENT VARIABLE. For example, suppose
you are trying to establish the relationship between consumers' income and their spending on your product. Your data include figures on
incomes and savings. In the short run at least there is likely to be a close
correlation (collinearity) between incomes and savings, which will
cloud your analysis. Eliminate such problems by:

◼ using only the more important of the two independent variables
  (eg, incomes);
◼ combining the two variables into one (eg, create a new series
  incomes minus savings); or
◼ finding a new variable which substitutes for them both (eg,
  consumer spending).

A high degree of correlation between several variables is MULTI-
COLLINEARITY.

Combination. The number of ways of selecting or arranging things
when duplication is not allowed and order is unimportant. When choosing x items from a set of n, there are [n! ÷ (n − x)!] ÷ x! combinations (n!
is n FACTORIAL). See also MULTIPLE (duplication allowed) and PERMU-
TATION (no duplication, but order is important). Figure 1.4 shows how
to identify the correct counting technique.

Completion time. Unless there is evidence of a better alternative, use the
BETA DISTRIBUTION (see Activity and project times box, page 171) to model
completion times for individual activites; use the NORMAL DISTRIBUTION
to model overall completion time for a project involving many activities.

Compound interest. Interest which itself earns interest; interest which is added to the PRINCIPAL (initial sum) to create a higher amount on which interest accrues in the next period. The FREQUENCY of compounding (the number of CONVERSION PERIODS in a year) can have a significant effect. For example, 10% a year compounded four times a year yields an EFFECTIVE INTEREST RATE of 10.38% and is therefore better than 10.25% a year compounded once a year.

Conditional probability. The PROBABILITY of an event when it is conditional on another event occurring. The probability that a group of consumers will spend more on your product may be conditional on a fall in interest rates.

Confidence. The degree of certainty in estimating a figure. When SAMPLE data only are available a descriptive measure such as an AVERAGE cannot be calculated exactly. ESTIMATION identifies it within a given margin of error (the confidence interval) at a given level of confidence (the confidence level). The boundaries of the confidence interval are called confidence limits. See INFERENCE.

Constant. Anything that is fixed for the purposes of analysis, such as price per unit, as opposed to VARIABLES such as the number of units sold.

Constant prices. Volumes quoted in prices ruling on a BASE date. See VALUE AND VOLUME.

Constant sum game. In GAME STRATEGY, a game where total winnings always equal a fixed amount. In a battle for market share, the players always win 100% of the market in aggregate.

Constraint. A limitation. Formalised in mathematical programming with INEQUALITIES such as h < 10 indicating, perhaps, that a machine will be available for less than ten hours.

Contingency test. A STATISTICAL TEST to see if two VARIABLES (eg, the age and sex of customers) are related or if they are independent, that is, a GOODNESS OF FIT TEST in disguise. Results are tabulated in contingency tables.

Continuous data. Information which is reported only according to the accuracy of the measuring instrument (eg, weight of chickens: 500gm, 500.5gm, 500.00001gm, etc). Compare with DISCRETE DATA (eg, number of chickens).

Conversion factors. See also SI UNITS.

Unit A	Unit B	To convert from A to B multiply by	To convert from B to A multiply by
Length			
Inches	Millimetres	**25.4**	0.03937
Feet	Metres	**0.3048**	3.28084
Yards	Metres	**0.9144**	1.093613
Miles (1,760 yd)	Kilometres	**1.609344**	0.621371
Area			
Inches^2	Millimetres^2	**645.16**	0.001550
Feet^2	Metres^2	0.092903	10.76391
Yards^2	Metres^2	0.836127	1.195990
Acres (4,840 yd^2)	Metres^2	4046.86	0.000247
Miles^2	Kilometres^2	2.58999	0.386102
Volume			
Inches^3	Centimetres^3	**16.387064**	0.061024
Feet^3	Metres^3	0.028317	35.31467
Yards^3	Metres^3	0.764555	1.307951
UK pints	Litres	0.568261	1.759754
US pints (liquid)	Litres	0.473176	2.113376
UK gallons	Litres	4.54609	0.219969
US gallons (231 in^3)	Litres	3.78541	0.264172
UK gallons	US gallons	1.200950	0.832674
US barrels (petroleum)	Litres	158.987	0.006290
Mass			
Tons (2,240 lb)	Tonnes (1,000 kg)	1.01605	0.984204
Cwt (112 lb)	Kilograms	50.8023	0.019684
Pounds (lb)	Kilograms	**0.45359237**	2.204623
Imperial ounces (oz)	Grams	28.3495	0.035274
Troy oz (apothecaries oz)	Grams	31.1035	0.032151
Fuel consumption			
UK: miles/gallon	Km/litre	0.354006	2.824811
US: miles/gallon	Km/litre	0.425144	2.352144
Km/litre	Litres/100km	Divide into 100	
Temperature			
Fahrenheit	Celsius/centigrade	$(F - 32) \times 5/9$	$(C \times 9/5) + 32$

Note: Numbers in bold type are exact conversion factors. Units are those in international use. For further information see www.NumbersGuide.com.

Conversion period. The length of time between interest payments. Critical for comparing COMPOUND INTEREST rates. Reduce all interest rates to EFFECTIVE INTEREST RATES (the equivalent with interest compounded once a year) for successful comparison.

Correlation. A measure of the strength of a relationship between two sets of data. Relationships are usually identified with REGRESSION ANALYSIS.

- ◪ The COEFFICIENT of correlation r takes a value between −1 and +1. The nearer r is to zero, the weaker the correlation (see Figure 5.7).
- ◪ The coefficient of determination r^2 takes a value between 0 and 1, indicating the PROPORTION of the relationship between two VARIABLES explained by the regression equation (not necessarily a causal link). If $r^2 = 1$ then 100% of the relationship is explained by the equation. If $r^2 = 0$, the technique used failed to find any link.

Interpret sensibly. A high figure may indicate that the two variables are linked to some unidentified third series, which is not very helpful. A low value may not mean "no correlation", but perhaps just that LINEAR REGRESSION failed to find a good relationship. Consider a TRANSFORMATION and look out for AUTOCORRELATION and SERIAL CORRELATION.

Cost of capital. The WEIGHTED-AVERAGE cost of financing business capital. For example, a highly geared company with $8m loan stock at 10% and $2m share capital at 15% has an overall cost of capital of 11%. This is an important yardstick for INVESTMENT ANALYSIS. The average return on investments must be greater than the cost of capital, or the company will make a loss.

Counting technique. A method for finding the number of COMBINATIONS, MULTIPLES or PERMUTATIONS for a decision, etc. Figure 1.4 shows how to identify the required technique and make the calculation.

Crashing. Speeding up a project by spending more. See CRITICAL PATH ANALYSIS.

Credit. Money you own, perhaps paid into (credited to) your bank account. The opposite is DEBIT.

Critical path analysis (CPA). The European equivalent of the US PERT. These are techniques for planning and managing large projects. Both identify the order of priorities for various activities, the timetable for all parts of the project, and the shortest possible overall completion time. CPA further identifies crash times: the shortest possible completion time for each activity if additional resources are made available. PERT incorporates PROBABILITIES that each activity will be completed on time. A combination of the two, with plenty of attention to costs, is more useful for business in general.

Critical value. The boundary between accepting and rejecting a hypothesis. Known as a CONFIDENCE limit in straightforward ESTIMATION. See INFERENCE.

Cross-impact matrix. A DECISION TABLE highlighting the PROBABILITIES of pairs of events occurring simultaneously. List all events both down the side and across the top. Fill in one row of probabilities, and the rest follow automatically. Such a table helps focus attention on RISKS.

Cross-sectional data. Information relating to one point in time (eg, sales by branch in June). Compare with TIME SERIES, like monthly sales.

Cryptology. See ENCRYPTION.

Currency. See EXCHANGE RATE.

Current prices. Actual prices, as opposed to CONSTANT PRICES. See also VALUE AND VOLUME and CURRENT VALUE.

Current ratio. A RATIO used to measure a company's liquidity: current assets divided by current liabilities. A higher ratio suggests greater liquidity, but it may also indicate too much cash tied up in non-productive assets such as cash or accounts receivable. There is no correct level; compare with other similar businesses. See also ACID TEST.

Current value. Today's spending power of a given sum available at a future date, after adjusting for expected INFLATION. See also PRESENT VALUE, which is the same arithmetic applied to interest rates. Current value and present value are two essential ingredients for INVESTMENT ANALYSIS.

Current-weighted index. A WEIGHTED AVERAGE in index form where the weights change over time. For example, consumer price indices are usually current-weighted to reflect changes in the pattern of consumption. It is complicated, difficult to keep up-to-date and understates long-run changes. Sometimes called a Paasche index after its founder. Compare with a BASE-WEIGHTED INDEX (or Laspeyres) where the weights are fixed.

Curvilinear relationship. A relationship which draws as a curved line on a chart. It can often be straightened by TRANSFORMATION. This allows techniques such as linear regression and LINEAR PROGRAMMING to be applied to curvilinear relationships.

Cycle. In a TIME SERIES, a pattern recurring less frequently than once a year (otherwise it is SEASONALITY). Identify cycles using annual data: there may be more than one (eg, a five-year business cycle and a ten-year product cycle). Find the TREND and the difference between the raw data and the trend is the cyclical component.

D

Debit. Money you owe, perhaps deducted (debited) from your bank account. The opposite is CREDIT.

Deciles. The values which split a set of ranked observations into ten equal parts. Sometimes the term is used to refer to the set of values between two deciles. See also PERCENTILES, QUARTILES and MEDIAN.

Decimal equivalents. Financiers often insist on quoting interest rates in eighths, sixteenths and thirty-seconds.

Decimal places. The number of digits to the right of a decimal point.

1.234 has three decimal places. Balance accuracy (many decimal places) with easy reading (few decimal places).

Fractions and their decimal equivalents

1/32	0.03125	9/32	0.28125	17/32	0.53125	25/32	0.78125
1/16	0.0625	5/16	0.3125	9/16	0.5625	13/16	0.8125
3/32	0.09375	11/32	0.34375	19/32	0.59375	27/32	0.84375
1/8	**0.125**	**3/8**	**0.375**	**5/8**	**0.625**	**7/8**	**0.875**
5/32	0.15625	13/32	0.40625	21/32	0.65625	29/32	0.90625
3/16	0.1875	7/16	0.4375	11/16	0.6875	15/16	0.9375
7/32	0.21875	15/32	0.46875	23/32	0.71875	31/32	0.96875
1/4	**0.25**	**1/2**	**0.5**	**3/4**	**0.75**		

Note: Fractions in bold are the more common ones.

Decision act or decision alternative. An option open to the decision-maker (eg, buy, hold, sell) as opposed to their OUTCOMES, which may depend on chance.

Decision analysis. QUANTITATIVE technique for making decisions. In essence it identifies all sub-decisions; forces a rigorous analysis of all possible DECISION ALTERNATIVES; quantifies the PROBABILITY of each OUTCOME; helps evaluate how much to spend on improving the inputs to the decision; and identifies the optimal decision.

Decision table. A tabulation of DECISION ALTERNATIVES and their OUTCOMES to help identify the optimal decision.

Decision tree. A diagrammatic representation of DECISION ALTERNATIVES and their OUTCOMES. Trees often become highly complicated, but they force rigorous analysis.

Decomposition. Breaking a TIME SERIES into its component parts: TREND, CYCLE, SEASONALITY and RESIDUAL. There may be more than one cycle (eg, a five-year business cycle and a ten-year product cycle). For simple forecasting, try to identify a trend from annual data and the seasonality from more frequent observations. Project the trend then add back the seasonal component.

Decryption. See ENCRYPTION.

Deduction. ESTIMATION of a SAMPLE statistic from POPULATION data. For example, if you know that the average weight of all students at a college (the population) is 160lb, you can deduce that ten of them (a sample) in an elevator are likely to weight 1,600lb in total. See INFERENCE.

Degeneracy. A situation at a certain stage of mathematical programming where there is no straightforward way of determining which of two variables should enter the analysis next. This may lead to cycling between two non-optimal solutions.

Delphi. A QUALITATIVE forecasting process which combines individual judgments into a consensus. The process tries to avoid the influence of personality and rank, but it is open to manipulation by individuals who maintain extreme views.

Dependent variable. A VARIABLE which depends on (is calculated from or projected by) other variables. Sales revenue is dependent on price and quantity (which are the INDEPENDENT VARIABLES). In REGRESSION ANALYSIS, the dependent variable is usually labelled y.

Depreciation. In accounting, an allowance for the fact that fixed assets (such as plant and machinery) wear out or become obsolete. For example, a $10,000 clump press with an expected life of five years and a scrap (residual) value of zero might be depreciated (charged against profits) at the rate of, say, $2,000 a year. Do not use this accounting concept for analysing the costs of production (see IMPUTED COSTS) or for providing for the replacement of the assets (see SINKING FUNDS).

Deprival value. An asset's value to a business; the amount by which a business would be worse off without the asset.

Descriptive measures. Shorthand summaries of data DISTRIBUTIONS. AVERAGES (such as the MEAN, MEDIAN or MODE) identify a distribution's mid-point and its location along the set of all numbers. Measures of SPREAD (eg, STANDARD DEVIATION or SEMI-INTERQUARTILE RANGE) describe how widely the distribution is scattered around the mid-point. SKEW and KURTOSIS indicate the shape of the distribution, although standard distributions are more useful.

Deterministic. Dependent on certainties rather than PROBABILITIES. The allocation of raw materials among production processes is deterministic if production requirements are defined, but most business problems are STOCHASTIC.

Deviation. Difference (see MEAN ABSOLUTE DEVIATION, STANDARD DEVIATION).

Discount. Money off. Popularly a reduction in price for an immediate cash payment. In finance, interest in advance. The discount rate (interest rate) is always misleading. Discounting can double the EFFECTIVE INTEREST RATE on consumer loans. For example, a loan of $1,000 for one year with 10% a year interest discounted in advance attracts interest of $100. This $100 is equivalent to an effective interest rate of nearer 20% if the interest is calculated on the (declining) outstanding balance.

Discounted cash flow. Technique for investment appraisal which converts all cash flows to a standard form for easy comparison. There are two approaches, NET PRESENT VALUE and INTERNAL RATE OF RETURN. See INVESTMENT ANALYSIS.

Discrete data. Information counted in steps (eg, number of employees). Compare with CONTINUOUS DATA, like heights of employees.

Discriminant analysis. A method of predicting into which group an item will fall (eg, creditworthy or non-creditworthy). A z SCORE (nothing to do with z and the NORMAL DISTRIBUTION) is calculated as a WEIGHTED AVERAGE of influences and used as the basis for the decision. For example, credit companies often determine credit limits from factors such as age, length of time as a home owner, occupation, and so on. Weights are fitted by observation and experience, which may occasionally throw up odd results.

Dispersion. See SPREAD.

Distribution. Any set of related values; eg, salaries paid by one firm or daily output of one machine. A distribution is described by an AVERAGE (eg, MEAN) which identifies its midpoint, a measure of SPREAD (eg, STANDARD DEVIATION), and a measure of its shape. Shape is summarised either by SKEW and KURTOSIS (peakedness) or, more helpfully, by a standard distribution. such as: the NORMAL DISTRIBUTION; the POISSON DISTRIBUTION; the EXPONENTIAL DISTRIBUTION; the BETA DISTRIBUTION; BINOMIAL.

Dominance. In GAME STRATEGY, eliminating a MATRIX row or column that would never be played because there are better PAYOFFS dominating elsewhere. Dominance may be used to reduce matrices to 2×2 format to simplify analysis.

Dual. The underlying value of a resource, such as the profit that is earned from adding one extra unit of a raw material. Known in economics as the shadow price. Useful for what-if SENSITIVITY ANALYSIS. See LINEAR PROGRAMMING.

Durbin-Watson coefficient of serial correlation. Measures SERIAL CORRELATION among RESIDUALS in REGRESSION ANALYSIS. DW will fall in the range 0–4, with 2 indicating a RANDOM series. If the residuals are not random, more analysis is required.

E

e. The constant 2.7182818… Used as a BASE for natural LOGARITHMS and in various calculations involving growth.

Econometrics. Mathematical methods applied to economic problems. The results need careful interpretation since it is seldom if ever possible to model all elements of capricious human behaviour. Results often depend heavily on RESIDUALS which are massaged subjectively by the econometrician. See also MODELLING.

Economic order quantity (EOQ). Optimal volume of stock to order in STOCK CONTROL. See REORDER POINT.

Effective interest rate. A yardstick for comparing interest rates. It is the rate of interest which would be paid if interest were compounded exactly once a year. If the annual rate of interest is r and the number of times it is paid in a year is k, the effective interest rate r_e is found from $(1 + r_e) = [1 + (r/k)]^k$. For example, with 10% a year compounded every three months: $(1 + r_e) = [1 + {}^{0.10}/4)]^4 = 1.025^4 = 1.1038$. In other words, 10% a year compounded every three months is equivalent to 10.38% a year compounded once a year. This is better than a rate of, say, 10.25% a year compounded once a year. US credit contracts are required to specify the effective interest rate. The UK equivalent is APR which also includes some loan fees and related charges. See also COMPOUND INTEREST.

Encryption. Encryption relies on mathematical algorithms to scramble text so that it appears to be complete gibberish to anyone without the *key* (essentially, a password), to unlock or decrypt it.

Equation. A numerical relationship such as profit = (sales × price) − costs. The equality = is the balancing point. Brackets indicate priority; perform operations in brackets first. See SOLVING EQUATIONS, pages 20ff.

Error. See ERROR OF COMMISSION, MEAN FORECAST ERROR, STANDARD ERROR.

Error of commission and error of omission. Two ways of making a mistake when doing a HYPOTHESIS TEST (see also INFERENCE). An error of commission, or type I error, is doing something in error, such as crossing a road when it is not safe to do so. An error of omission, or type II error, is doing nothing when something could be done, such as not crossing a road when it was actually safe to do so. In this clear cut case, several errors of omission are preferable to one error of commission. More commonly, business decision-makers have to try to balance the risks of each type of error.

Estimation. One way of decision-making when SAMPLE data are involved. In this case, a descriptive measure such as an AVERAGE cannot be calculated exactly. Estimation identifies it within a given margin of error (confidence interval) at a given level of CONFIDENCE (confidence level). For example, from a sample of 30 observations, the average daily output of a machine might be estimated at between 2,990 and 3,010 floppy disks at a 95% confidence level. The boundaries of the confidence interval (2,990 and 3,010) are confidence limits. The size of the sample, the confidence level and the confidence interval are linked. Define two, and the third is fixed automatically. A sample of five observations might give 100% confidence that average daily output is between 0 and 6,000 disks, or 50% confidence that output is between 2,000 and 4,000 items. Records from the entire life of the machine (a 100% sample) would give 100% certainty that average output was 3,003.2 floppy disks. See also INFERENCE.

Exchange rate. The price of one unit of currency in terms of another. "The exchange rate of the dollar against the euro is 1.05" and "The rate for the dollar per euro is 1.05" both mean $1.05 = €1. Once you are clear about the currency with a value of 1, exchange rate calculations become straightforward. Always use the rate applying on the date of a transaction. Using an average rate over a period of, say, one month when several transactions took place can produce wobbly results.

Expectation. Average outcome if an event is repeated many times. The expected PROPORTION of heads when tossing a fair coin is 0.5.

Expected monetary value (EMV). EXPECTED PAYOFF measured in cash terms.

Expected payoff (EP). The average outcome if a decision could be repeated many times. If there is a 0.4 PROBABILITY of making a $50m profit and a 0.6 probability of making a $10m loss, the expected payoff is ($50m × 0.4) + (−$10m × 0.6) = $14m. EP (also known as EXPECTED MONETARY VALUE) is used in DECISION ANALYSIS.

Expected value of perfect information (EVPI). The additional profit that would result from having foretold the future and made the optimal decisions, compared with the EXPECTED PAYOFF without a crystal ball. Do not spend more than the EVPI on acquiring additional data. Since information is usually far from perfect, see EXPECTED VALUE OF SAMPLE INFORMATION.

Expected value of sample information (EVSI). The incremental profit that would result from collecting and using SAMPLE information, compared with the EXPECTED PAYOFF without this information. EVSI is less than the EXPECTED VALUE OF PERFECT INFORMATION since sample information is far from perfect. Again, do not spend more than the EVSI on acquiring additional sample data.

Exponent. The little number in the air used to indicate how many times to multiply a number by itself. For example, the 3 in $2^3 = 2 \times 2 \times 2 = 8$. Also known as a POWER and a LOGARITHM.

Exponential distribution. A standard DISTRIBUTION which models constant growth rates over time. The negative exponential distribution is used to model service times (POISSON) in the analysis of QUEUEING. One figure unlocks a wealth of detail. For example, if the average service rate is three per hour, the exponential reveals that there is a 22% probability of any one service taking more than 30 minutes. Formulae are given in SOME QUEUEING ARITHMETIC, page 162.

Exponential smoothing. A delightfully simple way of projecting a TREND. As with most forecasting techniques it uses only historical observations and is slow to respond to changes in trends.

Extrapolation. Forecasting with a ruler. Projecting a TREND forward on an implicit or explicit hope that what happened in the past will happen in the future; it might not, of course.

F

F test. A statistical test used to compare VARIANCE (SPREAD) in two SAMPLES. (For example, is the variance of machine A's output different from the variance of machine B's output?) If the RATIO variance of sample A divided by variance of sample B is greater than the CRITICAL VALUE listed in F-tables, it is concluded that there is a significant difference between the variances and one machine has a greater range of variation in its output than the other. Named in honour of the mathematician Fisher, who developed ANALYSIS OF VARIANCE, based on the F-test.

Factor analysis. A method of reducing problems to a manageable size by combining many VARIABLES into single factors. Industrial demand for a product might be analysed in relation to the output of each industry sector (autos, bakeries, breweries, and so on). Or the output of all sectors might be aggregated into a single factor, such as total industrial production. This is a good way of dealing with COLLINEARITY, but it is sometimes difficult to interpret the factors after their analysis.

Factorial, !. A declining sequence of multiplications denoted by an exclamation mark (!). 4! is $4 \times 3 \times 2 \times 1 = 24$. Handy for COMBINATIONS and PERMUTATIONS.

Feasible region. In mathematical programming, the collection (set) of all possible solutions to a problem from which the optimal solution will be selected.

Finite population correction factor (FPCF). A simple adjustment to STANDARD ERRORS when the SAMPLE is larger than about 5% of the POPULATION size. For example, if drawing a sample from a population of 20 items, there is a 1 in 20 PROBABILITY of any item being selected on the first draw, a smaller 1 in 19 chance for the second item, and so on. The FPCF adjusts for this changing CONDITIONAL PROBABILITY, which is not important with large populations.

Fixed costs. Business costs that do not change in the short term (such as factory rents). Compare with VARIABLE COSTS. Fixed costs are only fixed over a certain range of production. For example, increasing output beyond a certain level may mean renting a new factory and hence a step-change in fixed costs. See IMPUTED COSTS.

Flow chart. Diagrammatic representation of an activity. Flow charts help to identify all possible outcomes and the correct sequence of operations.

Forecasting. Predicting the future. See TIME SERIES and always consider the effect of SEASONALITY and INFLATION. Projection might be by EXTRAPOLATION, CAUSAL FORECASTING or QUALITATIVE methods. With all forecasts you should monitor OUTTURNS against prediction and adjust the forecast accordingly to improve it next time around (see MEAN FORECAST ERROR).

Foreign exchange. See EXCHANGE RATE.

Frequency. An ordinary word which mathematicians use to make simple concepts sound complicated. A frequency polygon is a GRAPH of a DISTRIBUTION showing how many times each observation occurs (its frequency). Relative frequency is another name for PROBABILITY.

Future value. The amount to which a sum of money will grow if invested at a given rate of interest. For example, $100 (the PRESENT VALUE) invested for one year with 10% interest credited at the end of the year will grow to a future value of $110. See INTEREST RATE PROBLEMS.

G

Game strategy. Analysis for dealing with competitive situations such as pricing policy. It fails if opponents do not behave rationally, and may sometimes result in the worst of all worlds.

Gearing. UK name for leverage, which is the extent to which a business is financed by loans rather than shareholders' equity (precise definitions vary; one is long-term loans divided by shareholders' equity). The higher the proportion of loans, the higher the gearing. High gearing may be good for shareholders (a small increase in profits results in a relatively large increase in dividends), but it increases the riskiness of the business (ie, the company is heavily dependent on borrowing).

Geometric mean. An average, used with TIME SERIES which are growing/shrinking over time. Multiply numbers together and take the nth ROOT, where n is the number of observations. The geometric mean of 100, 110 and 130 is $(100 \times 110 \times 130)^{1/3} = 112.7$. See also MEAN.

Goal programming. See LINEAR PROGRAMMING.

Goodness of fit test. A STATISTICAL TEST with a sensible name. It compares the expected DISTRIBUTION of a SAMPLE with actual results to see if the sample conforms to expectations. It might be used, for example, to test whether the pattern of daily output or sales is normally distributed. See NORMAL DISTRIBUTION.

Graph. A pictorial representation of data. Chapter 4 gives hints for drawing and interpreting graphs. Chapter 5 shows how they are used for identifying relationships between series of data. Chapter 9 shows how they can be used to find the optimal solution to a problem.

Greek letters. Used for shorthand notation, often to indicate PARAMETERS for POPULATIONS (as opposed to statistics for SAMPLES, which are identified by equivalent Roman letters). See SYMBOLS for a selective list.

Growth. Simple to handle using PROPORTIONS. TIME SERIES which are growing at a steady rate plot as curved lines on graphs. Straighten them out for LINEAR REGRESSION ANALYSIS using TRANSFORMATIONS with LOGARITHMS.

H

Heuristic. Problem solving by creative thinking rather than by logical analysis. Looking sideways at a problem and coming up with a heuristic solution (rather than using a lengthy conventional approach) may pay dividends in saved time.

Histogram. A BAR CHART where the area (rather than just the height) of each bar is representative of the magnitudes illustrated. Histograms are important because they are one step on the path to a smooth line GRAPH which aids analysis. See, for example, the NORMAL DISTRIBUTION in Figure 3.2.

Hurwicz criterion. A method of incorporating judgment into a decision under uncertainty. Reviewed along with three other alternatives on page 139.

Hypothesis test. A statistical approach to testing a theory. It confirms or refutes a hypothesis that a descriptive measure for a POPULATION accords with a figure estimated from SAMPLE data (or vice versa). The logic of ESTIMATION applies. In classical hypothesis testing, the hypothesis is rejected if the descriptive measure calculated from sample data is outside a given margin from the hypothesised value. Estimation's confidence limits become significance limits or critical values. Instead of measuring the confidence of being right (say 95%), the tests identify the risk of being wrong (the significance level; 5%, which is 100 minus the confidence level). For example, a paperweight manufacturer might form the hypothesis that the average daily output of his machine is 3,120 paperweights. The MEAN output calculated from sample data would be compared with a pre-determined CRITICAL VALUE (or significance limit) and the hypothesis would be accepted only if there was no more than, say, a 5% risk of being wrong (a 5% level of significance). The sample size, the significance level and the significance limits are interrelated. See ERRORS OF COMMISSION and ERRORS OF OMISSION. An alternative approach is PROBABILITY VALUE, or P-value hypothesis testing, where estimation techniques are used to calculate the PROBABILITY that the hypothesis is correct. See also INFERENCE.

I

Imputed costs. Realistic, estimated costs; the amount a company gives up by not selling or leasing an asset. For example, suppose a press, purchased some years ago, will produce net revenue of $1,000 a year. Is it worth running the machine? The only realistic way of costing it is by estimating the amount that the company could raise by leasing or selling the machine. If this imputed cost is less than the net revenue, the company should run the press. If the imputed cost is above the net revenue, the company is better off leasing or selling the machine. Use imputed costs instead of the accounting concept of DEPRECIATION (which reflects historic costs but not the current OPPORTUNITY COSTS of running the machine). See ABSORPTION COSTING and MARGINAL ANALYSIS.

Independence test. See CONTINGENCY TEST.

Independent. Unaffected by another event. The x variables in MULTIPLE REGRESSION ANALYSIS are assumed to be independent of each other. If they are not, results can be distorted and difficult to interpret. In such instances use multivariate methods such as FACTOR ANALYSIS. PROBABILITIES vary depending on whether or not events are independent. The probability of drawing two specified playing cards from a pack is $\frac{1}{52} \times \frac{1}{52} = \frac{1}{2704}$ if the first card is replaced before the second draw; the events are independent. But if the first card is not replaced the CONDITIONAL PROBABILITY is $\frac{1}{52} \times \frac{1}{51} = \frac{1}{2652}$ since the second card is selected from 51; the events are dependent.

Independent variable. A VARIABLE which is used to estimate a DEPENDENT VARIABLE. Sales revenue depends on two independent variables: sales volume and price. In REGRESSION ANALYSIS, the independent variables are usually labelled x.

Index number. A number expressed as a PROPORTION of (ie, divided by a) BASE value; more usually converted into a PERCENTAGE (by multiplying by 100). For example, 20 and 30 with the first as the base value are 100 (= $\frac{20}{20} \times 100$) and 150 (= $\frac{30}{20} \times 100$) in index form. Series such as INFLATION where the raw values are irrelevant, misleading or distracting are often converted into indices. Watch for the convergence illusion, which reflects the fact that two index numbers with the same base will always meet at that base (see Figure 1.2). When several factors are combined (such as in an inflation index) the components form a WEIGHTED AVERAGE. Such an index may be a BASE-WEIGHTED or CURRENT-WEIGHTED INDEX. The choice of the weighting can make a big difference to what the index appears to show.

Induction. Estimating a POPULATION PARAMETER from SAMPLE data. See INFERENCE.

Inequality. Numerical relationship where the two sides are linked by a greater than >, greater than or equal to ≥, less than <, or less than or equal to ≤ sign. CONSTRAINTS in mathematical programming start life as inequalities.

Infeasible. A mathematical programming problem that has no solution, usually because of mutually exclusive CONSTRAINTS. A problem is infeasible if, for example, sales must exceed 20,000 units to cover operating costs but maximum output capacity is 10,000 units.

Inference. Decision-making when SAMPLE data are involved. A descriptive measure such as an AVERAGE can be determined exactly only from a complete set of data. When just a sample is available, the average cannot be calculated exactly. Either DEDUCTION or INDUCTION is necessary. There are two main approaches: ESTIMATION and HYPOTHESIS TEST.

Inflation. A general rise in the level of prices. The problem is that it is not uniform. Wages may rise at a different rate from raw materials, and so on. In general, convert all money into CURRENT VALUES before analysis (BUDGET, FORECASTING, etc) begins.

Integer programming. A complex form of LINEAR PROGRAMMING which deals with whole numbers.

Intercept. The point at which two lines cross on a GRAPH. Identify using EQUATION solving.

Interest rate problems. Potentially complex, but simple to solve with a few rules. Convert all rates to EFFECTIVE INTEREST RATES and compare these. The arithmetic for handling ANNUITIES will solve most problems involving streams of payments (SINKING FUNDS, mortgages, school fees, etc). If INFLATION is involved, convert all money into CURRENT VALUES before analysis. See also INVESTMENT ANALYSIS.

Internal rate of return (IRR). The rate of COMPOUND INTEREST an investment project would earn if it was, in effect, a bank deposit; ie, the interest rate which makes the stream of future income equal, in today's prices, to the cost of the project. An excellent criterion for assessing investments since PERCENTAGE rates are familiar and easy to compare with alternatives. But it may not be possible to calculate the IRR for CASH FLOWS that swing between positive and negative more than once. The alternative is to evaluate projects using NET PRESENT VALUE. See INVESTMENT ANALYSIS.

Interquartile range. The SPREAD of a set of numbers after eliminating the first and last 25% (ie, the first and last QUARTILES).

Interval data. Information which can be ranked by relative magnitude (eg, heights, weights, lengths). Compare with CATEGORICAL and ORDINAL DATA.

Investment analysis. The review and comparison of investment projects by converting them into standard form. Think of it in terms of "If I put the capital in a 'safe' deposit account at an EFFECTIVE INTEREST RATE of x% a year would that yield a better return than this project offers?". There are two approaches.

- INTERNAL RATE OF RETURN (IRR) reveals the equivalent interest rate generated by a project. Pick the project that generates the highest rate, so long as it is above your COST OF CAPITAL or whatever yardstick you use.
- NET PRESENT VALUE (NPV) determines what profit a project would generate at a predetermined rate of interest (perhaps your cost of capital). Pick the project with the highest NPV, but note that varying the interest rate yardstick may rank projects in different orders.

PC SPREADSHEETS produce both NPV and IRR with ease. IRR – probably the better of the two – is sometimes impossible to calculate if CASH FLOW swings between positive and negative more than once. Also known as DISCOUNTED CASH FLOW, this analysis conventionally neglects INFLATION. If projected inflation is not constant or if cash flows from competing projects are received at different times, convert all money into CURRENT VALUES before investment analysis. PAY-BACK PERIOD and simple return on investment may be brought into the analysis, but do not use them alone. Use a DISTRIBUTION such as the normal curve to model the effect of RISK. PORTFOLIO STRATEGY extends investment analysis to find the collection of investments with the highest returns and lowest risks.

J

Judgment. An essential input to decision-making. Usually the best decision is reached by a combination of judgment and numerical analysis.

Judgmental expected payoff (JEP). EXPECTED PAYOFF calculated by factoring judgment into decisions under uncertainty. See pages 139–41 for a comparison with three other decision-making criteria.

K

Kurtosis. An esoteric descriptive measure of the extent to which a DISTRIBUTION is flat (platykurtic) or pointed (leptokurtic).

L

Lag. To happen after a time lapse. An increase in consumer spending may lag (happen after) a reduction in taxes or an increase in wage rates, rather than being coincidental (happen at the same time) or leading (happen before) the changes. When analysing data, it is frequently necessary to lag a series (shift it back in time) to take account of leads and lags. See also LEAD.

Laspeyres index. See BASE-WEIGHTED INDEX.

Lead. To happen in advance. An increase in wage rates may lead an increase in consumer spending, rather than being coincidental or LAGGING the change. Leading indicators are often useful for business forecasting. Interest rates might be used as a leading indicator of demand for carpets: house-building may pick up after a fall in mortgage interest rates and brand new homes need new carpets. Identifying the lead time and the magnitude of the change in demand is the tricky bit. They may differ over time. REGRESSION ANALYSIS of interest rates and a lagged index of carpet sales may reveal a link.

Leasing. Payment of rent in exchange for the use of an asset (such as a car or other machinery). Compare with the cost of buying by using INVESTMENT ANALYSIS.

Leptokurtic. See KURTOSIS.

Leverage. US name for GEARING.

Linear programming. A numerical way of finding the optimal solution to a problem when there may be a CONSTRAINT, and alternatives and an objective (such as profits or costs) need to be maximised or minimised. Linear programming deals with a single objective and linear (straight-line) relationships. Often an approximation to a linear problem is easier than complicated enhancements such as non-linear programming, INTEGER PROGRAMMING and goal programming. For example, try to straighten a CURVILINEAR RELATIONSHIP using a TRANSFORMATION and then apply linear programming techniques.

Linear regression analysis. A method of identifying straight-line relationships between two sets of data. See REGRESSION ANALYSIS.

Linear relationship. A relationship which plots as a straight line on a graph. Most straightforward techniques such as LINEAR PROGRAMMING and linear regression assume straight-line relationships.

Logarithm. Another name for a POWER or EXPONENT. Ten raised to the power of 3 is $10^3 = 10 \times 10 \times 10 = 1{,}000$. 3 is the logarithm of 1,000 to the BASE 10. Logs are handy for flattening out GROWTH rates. See TRANSFORMATIONS.

M

Marginal analysis. Analysis at the edge. For example, it is worthwhile adding another unit of output if the marginal revenue from its sale will exceed the marginal cost of production. Excellent where the number of DECISION ALTERNATIVES and OUTCOMES is large.

Markov chain, Markov system, named after a Russian mathematician. Link between the present and the future. When the next state of an event depends on the present state, its probable state at any point in the future can be calculated. In other words, if future sales depend on current sales, the long-run market share can be identified. Life is never that simple, but Markov chains are helpful for basic analysis.

Matching. The convention that profit is based on revenues and expenses relating to the accounting period, rather than cash actually received and paid out. For example, a sale on credit is recorded at the date of the sale, not the date that the cash is received. This generally makes good sense. But for analysis of production costs use IMPUTED COSTS rather than DEPRECIATION.

Mathematical programming. See LINEAR PROGRAMMING.

Matrix. Nothing more than a tabular presentation with a regular shape (all rows are the same width and all columns are the same length.) Matrix arithmetic allows complex calculations to be performed simply. Uses for matrices include GAME STRATEGY, mathematical programming and MARKOV CHAINS. Page 169 shows how two matrices are multiplied together.

Maximax. An optimistic decision-making criterion which identifies the decision alternative with the best (maximum) of best possible (maximum) outcomes. It is compared with three other criteria on page 139.

Maximin. A pessimistic decision-making criterion which identifies the decision alternative with the best (maximum) of worst possible (minimum) outcomes. See page 139.

Maximisation. Finding the highest solution, such as maximising profits in mathematical programming.

Maximum flow. A self-evident optimising technique for NETWORKS. See also OPTIMISATION.

Mean. Short for arithmetic mean, an everyday AVERAGE; add numbers together and divide by the total number of values. For example, the mean of 2, 3, 4, and 11 is $(2 + 3 + 4 + 11) \div 4 = 20 \div 4 = 5$. This is the best general purpose measure of the mid-point of a set of numbers. It is widely understood, easy to calculate and it uses every value in a series. But it is distorted by OUTLIERS. Never average two averages, which will give equal weight to two possibly unequal sets of data; average from the raw data or create a WEIGHTED AVERAGE. Always try to use an average with a measure of SPREAD; the STANDARD DEVIATION goes with a mean. Also always ask how any presented average is calculated; it may instead be a GEOMETRIC MEAN, MEDIAN, MODE, or weighted average.

Mean absolute deviation (MAD). The AVERAGE difference between a set of numbers and their MEAN. Plus and minus signs in the differences are ignored; eg, values −5 (below the mean) and +5 (above the mean) are both treated as +5. Simply add absolute differences between each number and the mean and divide by the number of observations. For example, the mean absolute deviation of 1, 2, 3, 4, 5 is $(2 + 1 + 0 + 1 + 2) ÷ 5 = 1.2$. MAD is sometimes used as a descriptive measure of SPREAD, although STANDARD DEVIATION is better. MAD is more useful for analysing forecast OUTTURNS (see MEAN FORECAST ERROR).

Mean forecast error. There are three main approaches to analysing differences between forecast predictions and OUTTURNS. They are all designed to cope with the problem that an outturn below the predicted value will be exactly offset by an outturn the same amount above (ie, these two deviations from forecast will average zero). To illustrate the approaches, suppose that sales of $1,000, $3,000 and $5,000 are forecast and actual sales are $1,500, $3,000 and $4,500.

- Mean absolute deviation (MAD) simply averages absolute differences between predictions and outturns: $($500 + $0 + 500) ÷ 3 = 333.33. This reveals the average forecasting error, but it does not penalise big errors.
- Mean squared error (MSE) eliminates minus signs by squaring the differences; add squared differences and divide by the number of observations: $($500^2 + $0^2 + -$500^2$) ÷ 3 = $166,666.67$. This gives extra weight to OUTLIERS but reports the error in units which are hard to relate to.
- Mean absolute percentage error (MAPE) eliminates minus signs by ignoring them while also giving extra weight to outliers; add absolute percentage differences and divide by the number of observations: $(^{$1,500}/_{$1,000} + {}^{$3,000}/_{$3,000} + {}^{$4,500}/_{$5,000}) ÷ 3 = (50\% + 0\% + 10\%) ÷ 3 = 20\%$. This involves the most lengthy calculations (which is not a problem with PC SPREADSHEETS), but it reveals the average error as a PERCENTAGE which is familiar to business managers.

Median. An AVERAGE; the middle observation. Arrange values in order and identify the central one. If there is an even number of observations, add the two middle ones and divide by 2. For example, the median of 2, 3, 4, 12 is $(3 + 4) ÷ 2 = 3.5$. Unless there is a good reason for picking the median or MODE, you should use the MEAN. The median is most useful if the mean is heavily distorted by OUTLIERS or if the mean cannot be calculated (eg, with ORDINAL DATA, such as good/fair/poor). An employee might more sensibly aspire to the median wage, rather than the mean. Try to use the median with its associated measure of SPREAD: the INTERQUARTILE RANGE.

Minimisation. Finding the lowest solution, such as the minimum cost, in mathematical programming.

Minimum distance. OPTIMISATION technique for finding the shortest route from A to B in a NETWORK. It can be tricky, but computers are good at finding it. Also called shortest path or shortest route.

Minimum span. Another OPTIMISATION technique, this time for finding the shortest way of connecting several points in a NETWORK when the order is unimportant. Relatively easy, but requires many fingers. Also called shortest span.

Mixed strategy game. A game without a SADDLE POINT, where the optimal strategy is to make one move in a specific percentage of the time, and the other move in the remaining time. See GAME STRATEGY.

Mod, modulo. A mathematical operation which gives the remainder after division by a number (the modulus). For example, 46 mod 11 = 2. The modulus is 11 and the answer is 2; in English, 46 divided by 11 is 4 remainder 2.

Mode. An AVERAGE; the most frequently occurring value. The mode of the series 2, 2, 3, 3, 3, 4, 4, 5 is 3, since there are three 3s but no more than two of any other number. A series may be bi-modal or multi-modal. Unless there is a good reason for picking the mode, you should use the MEAN (or possibly the MEDIAN). The mode might be the best summary of, say, paint sales: average (modal) sales might be 1 litre cans of white gloss. The mode is the only average that can be used with CATEGORICAL DATA.

Modelling. Using numerical methods and relationships to describe real-life situations. The formula profits = (units sold times sale price) less costs is a simple model of profits. The NORMAL DISTRIBUTION is frequently used to model a range of probable sales. SIMULATION is a way of modelling random events.

Modulus. See MOD.

Monte Carlo simulation. SIMULATION using RANDOM NUMBERS.

Moving average. A series where the values themselves are averages (MEANS) of values from several consecutive time periods. Table 5.1 shows how to calculate moving averages. They smooth out short-term fluctuations in a TIME SERIES and highlight whether any value is above or below the TREND.

Multicollinearity. A high degree of CORRELATION between several VARIABLES. See COLLINEARITY.

Multiple. The number of ways a decision can be made when duplication is permitted and order is important. For an identification code comprising two letters and a number, there are 26 letters (A–Z) that can go in the first position, 26 letters for the second, and 10 digits (0–9) for the third; there are $26 \times 26 \times 10 = 6,760$ multiples. See also PERMUTATION (duplication not allowed) and COMBINATION (order not important). Figure 1.4 shows how to identify the required counting technique.

Multiple regression analysis. A method of identifying relationships between several sets of data. See REGRESSION ANALYSIS.

Multistage sampling. A SAMPLING short cut, where the target sample is narrowed down in successive steps. A distribution company wanting to survey customer opinion could limit the cost of the survey this way. The company might first restrict the survey to ten of its 30 regions. Within those ten regions, it might select five areas. Within those areas, it might select ten offices, and so on until clusters of 100 customers were identified. Other techniques are QUOTA and STRATIFIED SAMPLING. See also INFERENCE, CLUSTER SAMPLING.

Multivariate analysis. Ways of handling many VARIABLES. See, for example, CLUSTER ANALYSIS, DISCRIMINANT ANALYSIS, FACTOR ANALYSIS and MULTIPLE REGRESSION ANALYSIS.

N

Net book value. Written down value; the historical cost of an asset (such as a machine) less ACCUMULATED DEPRECIATION. This is an accounting concept. Use instead IMPUTED COST for analysis of production costs, etc.

Net present value (NPV). The profit generated by a project in today's money: it is PRESENT VALUE of income less present value of outlays. Classical analysis does not allow for INFLATION; so you should calculate NPV from CURRENT VALUES. But note that a project which generates the highest NPV at one interest rate may not be the best at another interest rate. See INVESTMENT ANALYSIS.

Network. A visual representation of a project or system (such as a road map). See CPA/PERT (project planning), MINIMUM DISTANCE, MINIMUM SPAN and MAXIMUM FLOW.

Non-parametric method. A statistical decision-making method which does not require a knowledge of the DISTRIBUTION of a SAMPLING statistic and where data are on an ORDINAL scale. Non-parametric tests (such as the RUNS TEST) are less powerful than PARAMETRIC METHODS but often relatively simple.

Normal distribution. The most useful standard DISTRIBUTION. It describes many situations where the observations are spread symmetrically around the MEAN and applies where there are a large number of small influences at work. This includes many natural phenomena (such as IQs, crop yields, heights of trees) and many business situations (such as variations in items coming off a production line, expected sales and RISK in general). The mean and STANDARD DEVIATION unlock everything there is to know about any normal distribution. For example, 68% of values lie within ±1 standard deviation of the mean and 95% are within ±2 standard deviations. Normal tables (Table 3.2) allow any such percentages to be identified with Z SCORES.

Notation. Mathematical shorthand, such as $x + y = z$. It is initially off-putting, but it is very important to be familiar with. Pages 17–21 outline some basic notation. See also SYMBOLS.

Numerical programming. See LINEAR PROGRAMMING.

O

Objective function. The relationship to be optimised in mathematical programming, such as profits = x + y.

One-tail test. A decision-making process which analyses variation in one direction only. A packer guaranteeing minimum contents for bags of sweets might take a sample and test one tail of the DISTRIBUTION only. In contrast, a packer guaranteeing AVERAGE contents is interested in both tails since a bag with too many or too few items will be costly one way or another. This will require a TWO-TAIL TEST to make sure that the counting machine is accurate. See SAMPLING.

Operational research. Scientific methods applied initially to military operations and now to many disciplines including management. Techniques such as PERT were developed by the military.

Opportunity costs. An economist's method of valuing an activity: the cost of the best alternative forgone. The opportunity cost of the CAPITAL for a project is the income that could be generated, say, by putting the cash in an interest-bearing deposit account (see INVESTMENT ANALYSIS). The opportunity cost of using a machine is the revenue that could be raised by selling or leasing the machine (see IMPUTED COSTS).

Opportunity loss. The loss from not picking the best DECISION ALTERNATIVE. The minimum expected opportunity loss equals the EXPECTED VALUE OF PERFECT INFORMATION.

Optimisation. Finding the best answer to a problem. LINEAR PROGRAMMING does this. In contrast, SIMULATION identifies the best solution from only those tested.

Ordinal data. Information which can be classified into ordered categories where relative magnitudes are not identified (eg, poor/fair/good). Also known as ranked data. Compare with CATEGORICAL and INTERVAL DATA.

Outcome. The result of a decision. For example, when deciding where to open a new branch you would assess the range of potential outcomes (reflecting the success or otherwise of your marketing, the economic climate, and so on) for each site, and pick the site with the best set of outcomes. See DECISION ANALYSIS.

Outlier. An extreme value in a set of observations which may distort descriptive measures such as the MEAN or the RANGE. In the following set of numbers, 25 is the outlier: 1, 2, 3, 4, 25.

Outturn. Forecast-speak for what actually happened. Always compare expectations with outturn to see where a forecast went wrong and how it could be improved next time. See MEAN FORECAST ERROR.

P

Paasche index. See CURRENT-WEIGHTED INDEX.

Parameter. A descriptive measure relating to a POPULATION. The same measures calculated from SAMPLE data are known as statistics.

Parametric method. Statistical decision-making (such as the t-TEST and Z SCORE test) where the DISTRIBUTION of the SAMPLING statistic is known and data are on an interval or ordinal scale. Such tests produce more reliable results than NON-PARAMETRIC METHODS.

Payback period. Length of time required to recover investment outlay. Disregards pattern of flows and what happens after the outlay is recovered. Never use this criterion alone for judging capital projects. See also INVESTMENT ANALYSIS.

Payoff. Returns (cash, goodwill, manpower savings) resulting from a decision. DECISION ANALYSIS deals with EXPECTED PAYOFFS. GAME STRATEGY analyses decisions in competitive situations using payoff MATRICES.

PC spreadsheet. An essential tool for anyone dealing with numbers. See SPREADSHEET.

Percentage. A PROPORTION ($6 \div 50 = 0.12$) multiplied by 100 ($0.12 \times 100 = 12\%$). In other words, 6 is 12% of 50.

Percentiles. The values which split a set of ranked observations into 100 equal parts. Sometimes the term is used to refer to the values between two percentiles. See also DECILES, QUARTILES and the MEDIAN.

Permutation. The number of ways of selecting or arranging things when order is important and duplication is not permitted. When choosing x items from a set of n, there are n! ÷ (n − x)! permutations. (n! is n FACTORIAL). See also MULTIPLE and COMBINATION. Figure 1.4 shows how to identify the required counting technique.

Pi, π. Greek lowercase p. Used in statistics as shorthand for a PROPORTION of a POPULATION. More generally known as the CONSTANT 3.1415927... which is the ratio between the perimeter and the radius of a circle. See AREA AND VOLUME for practical applications.

Pie chart. A pictorial presentation where breakdowns are indicated by relative magnitudes of slices of a pie. Make sure that a comparison of two pies is a comparison of areas, not widths. Doubling the diameter quadruples the area. The area of a circle is $\pi \times r^2$, where π is the constant 3.1415927... and r is the radius.

Platykurtic. See KURTOSIS.

Poisson distribution. A standard DISTRIBUTION which models isolated events over time, such as arrivals of customers or telephone calls, or the incidence of accidents or mechanical failures. One figure unlocks a wealth of information. For example, if the average arrival rate of incoming telephone calls is 10 an hour, there is an 11% probability of 8 calls in any one hour and a 7.5% chance of a 15–30 minute delay between two calls. Formulae are shown on page 162. The poisson is frequently used with the EXPONENTIAL DISTRIBUTION in analysis of QUEUEING.

Population. All items in a defined group (all plastic bags manufactured today, all customers, etc) from which a SAMPLE might be drawn.

Portfolio strategy. An extension of classical INVESTMENT ANALYSIS to deal with RISK. The objective is to put together the collection of investments with the lowest overall risks and highest returns. If the chances of an increase or decrease in, say, equity prices are considered about equal, the STANDARD DEVIATION and the NORMAL DISTRIBUTION may be used to model risk. The problem with financial instruments is that the risks tend to be highly correlated (if one share decreases in price, others could well move in the same direction) so adding shares to the portfolio may increase rather than decrease the overall risk. The mathematics become quite complicated.

Posterior probability. A revised assessment. After conducting market research, a company might calculate posterior probabilities of its product succeeding by combining its PRIOR PROBABILITIES with the results of the survey. Such assessments provide a vital input to DECISION ANALYSIS.

Power (I). A number multiplied by itself a given number of times. Powers are essential for dealing with GROWTH rates and COMPOUND INTEREST. See also LOGARITHM.

Power (II). The effectiveness of a STATISTICAL TEST. If, for example, there is a 0.6 (60%) risk of an ERROR OF OMISSION in a particular test, the power of that test is $1 - 0.6 = 0.4$.

Prefix. Standard (SI UNIT) prefixes are listed in the table. In common currency are the terms million (10^6, or 1 and six zeros, 1,000,000) and billion (10^9, or 1 and nine zeros) although very occasionally a traditional British billion (10^{12}, 1 and 12 zeros) pops up.

Standard (SI unit) prefixes

Factor	Name	Symbol	Factor	Name	Symbol
10^{18}	exa	E	10^{-1}	deci	d
10^{15}	peta	P	10^{-2}	centi	c
10^{12}	tera	T	10^{-3}	milli	m
10^9	giga	G	10^{-6}	micro	m
10^6	mega	M	10^{-12}	namo	n
10^3	kilo	k	10^{-15}	pico	p
10^2	hecto	h	10^{-15}	femto	f
10^1	deca	da	10^{-18}	atto	a

Present value. The sum of money that would have to be placed on deposit today at a given COMPOUND INTEREST rate to generate a given sum in the future (the FUTURE VALUE). Present value and future value are linked by the interest rate (and any payments to or from investment). This is the basis for one approach to INVESTMENT ANALYSIS (the other approaches are NET PRESENT VALUE and INTERNAL RATE OF RETURN). Compare with CURRENT VALUE.

Prices. See VALUE AND VOLUME.

Price deflator. An index of INFLATION. For example, in national accounts statistics, consumers' expenditure and government consumption are subject to different rates of price increase; each is revalued from current to CONSTANT PRICES by dividing by the respective deflator. (Government spending suffers from what is known as the relative price effect, which means that suppliers know they can bump up the prices they charge the public sector by more than the man in the street would stand for.) See VALUE AND VOLUME.

Price/earnings ratio (P/E ratio). The market price of a share divided by earnings per share (EPS – the most recent figure for profits available to shareholders, divided by the number of ordinary shares). The higher the P/E ratio, the more highly thought of the company. Investors are prepared to bid up the price of the shares in the expectation of higher profits in years to come.

Principal. The capital sum placed on deposit or received as a loan, as opposed to the interest element. See INTEREST RATE PROBLEMS and INVESTMENT ANALYSIS.

Prime number. A positive integer greater than 1 that can be divided evenly only by 1 and itself. The first ten prime numbers are 2, 3, 5, 7, 11, 13, 17, 19, 23, and 29. The largest prime number known (as at January 2003) has over 4 million digits.

Prior probability. An initial assessment. Before conducting research, a company might assess the prior probability of its product being popular. After the survey, the revised probabilities are known as POSTERIOR PROBABILITIES. These assessments provide a vital input to DECISION ANALYSIS.

Probability. The likelihood of something happening expressed as a PROPORTION where 0 = no chance and 1 = dead certainty. The probability of an event is equal to the number of outcomes where it happens divided by the total number of possible outcomes. Probabilities may be determined as follows.

- A priori (ie, by logic): the probability of rolling a two with a six-sided dice is 1 ÷ 6 = 0.17.
- Empirically (by observation): if 1 in 100 items produced by a machine is known to be defective, the probability that a randomly selected item is defective is 1 ÷ 100 = 0.01.
- Subjectively (by judgment): in many instances, managers must use their experience to judge a likely outcome; eg, if we do this, there is a 0.3 probability that our competitors will produce a similar product within 12 months.

You can identify most probabilities by using the rules together with the COUNTING TECHNIQUES in Figure 1.4 and the BINOMIAL (page 125). See also PROBABILITY DISTRIBUTION.

Probability distribution. A precise name for any DISTRIBUTION used to model likely but as yet unobserved outcomes. The NORMAL DISTRIBUTION is used for describing observed measurements and, for example, probable levels of sales. The POISSON DISTRIBUTION is used as a probability distribution in modelling arrivals in QUEUEING.

Probability value, or P-value test. A STATISTICAL TEST of a hypothesis: the use of ESTIMATION techniques to calculate the probability that a hypothesis is correct. See also INFERENCE.

Profit margin on sales. A RATIO used in company analysis; trading profits divided by sales expressed as a PERCENTAGE. A low figure (perhaps 3%) may reflect poor performance, or perhaps high-volume low-margin activities such as food retailing. A higher figure may signal excess profits which will attract other companies to compete in the same market.

Program analysis and review technique (PERT). See CRITICAL PATH ANALYSIS.

Project planning. See CRITICAL PATH ANALYSIS.

Proportion. One number relative to (divided by) another: 6 is 0.06 of 100 (6 ÷ 100 = 0.06); easy to work with.

Pure strategy game. In GAME STRATEGY, a game in which each player has an optimal move which does not depend on prior knowledge of the opponent's play. A pure strategy game is identified by a SADDLE POINT.

Q

Qualitative. Judgmental, as opposed to QUANTITATIVE methods. Qualitative input is essential for business decision-making, and increases in value the better you know your business. Qualitative techniques which attempt to make judgment more rigorous, include CATASTROPHE THEORY, CROSS-IMPACT MATRICES, DELPHI forecasting and RELEVANCE TREES.

Quality control. The application of SAMPLING techniques to maintain production standards.

Quantitative. Scientific, as opposed to QUALITATIVE methods. Most techniques in this book are quantitative, although they attempt to take account of qualitative factors.

Quartile deviation. See SEMI-INTERQUARTILE RANGE.

Quartiles. The values which split a ranked set of data into four equal parts. Sometimes the term is used to refer to all the observations between two quartiles. The second quartile is better known as the MEDIAN. See also PERCENTILES and DECILES.

Queueing. An analysis of waiting lines which may be applied to queues of people at service counters, telephone calls arriving at switchboards, trucks arriving for deliveries or collections, and so on. Unless some other DISTRIBUTION is appropriate, arrivals are modelled by the POISSON DISTRIBUTION and service by the EXPONENTIAL DISTRIBUTION.

Quick ratio. See ACID TEST.

Quota sampling. A SAMPLING technique which requires careful handling. An interviewer might be asked to collect opinions from quotas of so many women aged 15–20, so many aged 21–30, and so on, and from similar quotas of men. This avoids such bias as might occur if the interviewer spoke only to good-looking or well-heeled men or women. Other sampling techniques are CLUSTER, MULTISTAGE and STRATIFIED SAMPLING. See also INFERENCE.

R

r and r². See CORRELATION.

Random. Without pattern or order. An essential concept for decision-making techniques. Tests for randomness include the RUNS TEST.

- PROBABILITY depends on random occurrence of events.
- Analysis of data is complete once the RESIDUALS are randomly distributed.
- SAMPLING theory depends on random selection of the sample.

Random numbers. A sequence of numbers generated randomly by drawing numbered balls from a hat or by a computer program or SPREADSHEET. Random numbers are used to model real-life events in SIMULATION.

Range. The distance between the lowest and highest values in a set of data; it can be distorted by OUTLIERS. Better measures of SPREAD are the SEMI-INTERQUARTILE RANGE (use with a MEDIAN) and STANDARD DEVIATION (use with a MEAN).

Ranked data. See ORDINAL DATA.

Rate of return. In INVESTMENT ANALYSIS, never judge a project by the simple accountant's rate of return (total returns divided by outlay times 100) which ignores the pattern of receipts over time. Instead use the INTERNAL RATE OF RETURN. See also RETURN ON CAPITAL EMPLOYED.

Ratio. Relationship used as a guide to company performance. Assess liquidity using the ACID TEST and CURRENT RATIO. Assess performance using PROFIT MARGIN ON SALES and RETURN ON CAPITAL EMPLOYED. Find out what the market thinks of the company by looking at its PRICE/EARNINGS RATIO.

Real value. Volume measured in prices ruling on a particular date. See VALUE AND VOLUME.

Regression analysis. A simple technique with an off-putting name for identifying relationships between data in two series (simple regression) or several series (multiple regression). A firm might use it, for example, to find the link between consumers' incomes and their spending on its goods or services. Or it might regress sales of its product against time (eg, months 1, 2, 3) so that it can forecast sales in a month yet to come (eg, month 48). SPREADSHEETS perform regression analysis with ease and always come up with a relationship. Use the correlation coefficient to test the strength of that relationship. But remember that a numerical relationship reveals nothing about cause and effect. Two series might be highly correlated due to a common link to a third, unidentified series. This is particularly true of things which are growing over time, perhaps due to INFLATION. The prices of hotel rooms in Paris and school text books may be rising at the same rate, but there is no direct relationship between them even though regression might suggest that there is. Also, poor correlation may reflect the wrong assumptions. For example, most PC regression programmes look for linear relationships only. Convert CURVILINEAR RELATIONSHIPS into straight lines by TRANSFORMATIONS. Chapter 5 gives hints for interpreting the results of regression analysis.

Relevance tree. A QUALITATIVE method for setting priorities. Start from the objective. Identify sub-objectives and allocate them weights according to their relevance or importance. Work backwards from each sub-objective until you reach the present. The tree will be a systematic framework for action, with weights indicating various priorities.

Reorder point, reorder quantity. The time to order a certain volume of stock. See STOCK CONTROL.

Residual. The amount unexplained by any analysis. TIME SERIES decomposition can identify CYCLES, TRENDS and SEASONALITY. The unexplained fluctuations in the series are the residuals. Similarly, REGRESSION ANALYSIS identifies relationships between two or more series. The residuals are the differences between the actual values and those calculated with the regression equation. Always examine the residuals for a pattern. Only when there is no pattern remaining can you conclude that no further analysis is possible.

Return on capital employed. A key RATIO used in company analysis: trading profits divided by capital employed, then expressed as a percentage. A low figure implies low profitability and perhaps an inability to ride out a recession. A high figure makes the company an attractive target for a takeover.

Risk. Uncertainty quantified with (judgmental) PROBABILITIES. Personal attitude to risk may be summarised by UTILITY (the value you attach to a particular sum of money). Risk in business decisions is often modelled by a DISTRIBUTION such as the NORMAL DISTRIBUTION (see also Z SCORE). The aim is to minimise (or at least take account of) risk as measured by STANDARD DEVIATION.

Roll-back. The process of starting at the end of a DECISION TREE and working back towards the first DECISION ALTERNATIVE, identifying PROBABILITIES and EXPECTED PAYOFFS along the route.

Root. The opposite of a POWER: the factors of a number which when multiplied together n times produce that number. This n may take any value and the root is known as the nth root. For example, the second root of 9 is 3 ($3 \times 3 = 9$), the fourth root of 625 is 5 ($5 \times 5 \times 5 \times 5 = 625$). The yth root of x is symbolised by $\sqrt[y]{x}$ or $x^{1/y}$. Most calculators have a $x^{1/y}$ key which makes taking (calculating) roots as easy as multiplication. They are useful for dealing with GROWTH rates.

Runs test. A STATISTICAL TEST for randomness. A run is a sequence of similar events. For example, this sequence from tossing a coin (HH TTT HH TT HHH) has five runs. A certain number of runs of certain lengths is predicted by PROBABILITY. If the pattern does not follow expectations, a non-random element has crept in.

S

Saddle point. In GAME STRATEGY, the PAYOFF which is lowest in its row and largest in its column. Such a payoff, known as the value of the game, identifies it as STRICTLY DETERMINED.

Sample. A subset of a POPULATION: perhaps 50 customers used to estimate the collective opinions of all customers.

Sampling. Selecting a SAMPLE and using it to make INFERENCES about the POPULATION from which it was drawn. The inferences will be false unless the sample is selected in an unbiased, RANDOM way. Sampling techniques include CLUSTER, MULTISTAGE, QUOTA and STRATIFIED SAMPLING.

Scatter. See SPREAD.

Scatter graph. A pictorial presentation where the points are not joined together. Very useful for identifying relationships (CORRELATION) between two sets of data (see Figure 4.6).

Seasonality. In a TIME SERIES, a pattern repeating itself at the same time every year reflecting seasonal factors such as sales of garden goods in the spring and sales of winter clothes in the autumn. Identify a seasonal factor for, say, January by finding the MEAN difference between each January figure and the MOVING AVERAGE. Or short-cut seasonality by examining changes over 12 months to compare like with like. Note that neither method catches year-to-year quirks such as the variable number of shopping days in a given month or the date of Easter.

Semi-interquartile range. Enthusiasts also know this as the quartile deviation. A measure of SPREAD which is one-half of the middle 50% of the distribution. If the first quartile is 14.5 and the third is 27.5, the semi-interquartile range is $(27.5 - 14.5) \div 2 = 6.5$. It cuts off extreme values and works well with skewed distributions, but since it uses only two observations, it is usually better to use the mean and STANDARD DEVIATION.

Sensitivity analysis. What-if analysis where a model is rerun to see what would happen if any particular figure is changed (eg, what if sales were 2% lower). It is easily done with SPREADSHEETS, but remember that it only shows what-if for the values you try.

Serial correlation. A non-random pattern in RESIDUALS, where each one is correlated with the previous one. The DURBIN-WATSON statistic helps identify this undesirable effect.

Shadow price. See DUAL.

Shortest path or shortest route. See MINIMUM DISTANCE.

Shortest span. See MINIMUM SPAN.

SI units. Système International d'Unités (International System of Units) agreed by the General Conference on Weights and Measures in 1961: a set of standards beloved of scientists, adopted by metric countries and growing in popularity. (See also CONVERSION FACTORS and PREFIX.) The SI system is based on seven base units (see table), although it enshrines a mass of other terminology.

SI base units

Quantity	Unit name	Symbol
Length	metre	m
Mass	kilogram	kg
Time	second	s
Electric current	ampère	A
Thermodynamic temperature	kelvin	K
Luminous intensity	candela	cd
Amount of substance	mole	mol

Significance test. A statistical approach to testing a theory. Significance tests indicate whether a descriptive measure (such as an average or proportion) calculated from SAMPLE data is representative of that for a population as a whole. See INFERENCE.

Significant figures. The number of digits with precise values. 1,062.6 (5 significant figures) rounds to 1,100 to two significant figures or 1,000 to one significant figure. Three or four is usually adequate for presentations, where increased accuracy can confuse the audience and detract from the point you want to make.

Simple interest. Rent for the use of money, paid solely on the PRINCIPAL and not compounded (see COMPOUND INTEREST).

Simplex algorithm. A set of rules for solving LINEAR PROGRAMMING problems. Tedious but simple to use and extremely powerful.

Simulation. The use of a model to represent real life; numerical simulation allows low-cost risk-free testing of, say, a QUEUEING or STOCK CONTROL system. The exercise can be repeated many times and rarely encountered events (such as a 1930s-style depression) can be modelled in. One approach is to use RANDOM numbers to pick points on a PROBABILITY DISTRIBUTION which models events such as sales. PC

SPREADSHEETS make this easy. Note that SIMULATION will not optimise, it will only reveal the best of those plans which are tried out.

Sinking fund. Cash reserve set aside to provide for replacement of, for example, a machine which is wearing out.

Situation. Uncontrollable influence in DECISION ANALYSIS. Also called states of nature and uncontrollables.

Skew. A descriptive measure of lopsidedness in a DISTRIBUTION. The distribution of wages and salaries is usually skewed. Most wages are clustered around the MEAN, with the gap between the lowest and the mean being much smaller than the distance from the mean to the highest. On a GRAPH, such a distribution would have a hump on the left with a longer tail to the right, so it is identified as positively skewed or skewed to the right. The opposite is negative skew, or skewed to the left. An unskewed distribution is symmetrical about the mean, as is the NORMAL DISTRIBUTION.

Slack. Spare resources. In project planning, spare time in one activity which might be used elsewhere (see CRITICAL PATH ANALYSIS). In mathematical programming, the situation where a resource is not fully used up in the optimal solution. A workshop available for ten hours a day might be in use for eight hours. The two hours of slack time could be allocated to another activity. If the resource is fully used up, it is said to be tight.

Spread. The range of values in a DISTRIBUTION, also known as scatter and dispersion. Spread measured by the difference between the highest and lowest values may be distorted by OUTLIERS. STANDARD DEVIATION (use with a MEAN) and SEMI-INTERQUARTILE RANGE (use with a MEDIAN) are preferable.

Spreadsheet. A grid for tabulating data. PC spreadsheets make repeated calculations simple, aid accuracy, can be larger than pen and paper would readily permit, allow easy what-if SENSITIVITY ANALYSIS and provide many tools for analysis. For example, they open up GRAPH drawing, REGRESSION ANALYSIS and INTERNAL RATE OF RETURN calculations to the masses. Such "automated" analysis needs to be interpreted carefully.

Spurious correlation. A relationship which appears to be good statistically but is not in reality. For example, two series (alcohol consumption and bankruptcies) might appear to be related when they are actually both linked to some unidentified third series such as a slump in national output.

Standard deviation. A measure of the SPREAD of a set of numbers. It is tedious to compute without a spreadsheet, but it makes a MEAN meaningful. See NORMAL DISTRIBUTION for hints on interpreting the standard deviation and Z SCORE for an easy way of estimating standard deviation. Among many other things, it is useful in business for MODELLING risks.

Standard error. The STANDARD DEVIATION of an estimate based on SAMPLE data. The standard error is used to calculate CONFIDENCE limits for the estimate. For example, a REGRESSION ANALYSIS that predicted sales at 10,000 units with a standard error of 50 units is projecting sales at 10,000 ± 100 (two standard errors) at a 95% confidence level. See NORMAL DISTRIBUTION for hints on interpreting standard deviations and, therefore, standard errors.

States of nature. See SITUATION.

Stationary. Not moving; no TREND. Demand for canteen lunches may vary from day to day but there may be no long-run increase or decrease in the level of fluctuation. Such stationary data may be easily projected using MOVING AVERAGES or EXPONENTIAL SMOOTHING.

Statistical test. Off-the-shelf procedures to quantify and test INFERENCES about descriptive measures calculated from SAMPLE data. PARAMETRIC TESTS (such as the t-TEST and Z SCORE TEST) are preferred as they are more powerful. If the DISTRIBUTION of the sampling statistic is unknown or data are ordinal, non-parametric tests (such as the RUNS TEST) must be used.

Stochastic. A problem involving PROBABILITIES, as opposed to a DETERMINISTIC problem based on certainties.

Stock control. A balancing act to minimise the risk of running out of stock against the cost of carrying too much. The review of stock control (pages 164ff) indicates ways of identifying the economic order quantity (REORDER QUANTITY) and the REORDER POINT, and touches on just-in-time stock control. Page 156 shows how a DISTRIBUTION such as the normal is used to identify the optimal stockholding.

Stratified sampling. A SAMPLING short-cut which requires careful handling. Where a POPULATION such as factory employees contains 80% women, a stratified sample might include a similar proportion, say, 80 females and 20 males chosen randomly from their respective sub-populations. Other techniques are CLUSTER, MULTISTAGE and QUOTA SAMPLING. See also INFERENCE.

Strictly determined. In GAME STRATEGY, a game in which both players have a pure strategy and neither can improve their PAYOFF by prior knowledge of the other's move. Such a game is identified by a SADDLE POINT.

Summary measure. A descriptive measure, such as a MEAN, PROPORTION, VARIANCE or STANDARD DEVIATION.

Symbols. Here is a list of the more common ones (see pages 18–20):
Operators
+ Addition: $2 + 3 = 5$
− Subtraction: $6 − 3 = 3$
× Multiplication, though sometimes the sign is omitted altogether:
$3 \times 2 = 3 \cdot 2 = (3)(2) = 6.$
÷ Division, written in various ways: $6 \div 2 = \frac{6}{2} = {}^6\!/_2 = 6/2 = 3$
x^y x raised to the POWER of y: $5^3 = 5 \times 5 \times 5 = 125$
${}^y\sqrt{x}$ or $x^{1/y}$. The yth ROOT of x: ${}^3\sqrt{125} = 125^{1/3} = 5$
! factorial. A declining sequence of multiplications: 4! is $4 \times 3 \times 2 \times 1 = 24$; handy for PERMUTATIONS and COMBINATIONS.
Σ Sigma, uppercase Greek S. Shorthand for take the sum of.
Other symbols
= equals: $6 + 3 = 7 + 2$
≈ approximately equals: $2.001 \approx 2$
≠ does not equal: $6 \neq 7$
> greater than: $7 > 6$
≤ less than or equal to
< less than: $6 < 7$
≥ greater than or equal to

( ) { } [ ] Brackets. Used to signify orders of priority. For example:

$(2 + 3) \times 4 = 20$

Greek letters

α alpha, lowercase a. Shorthand for ERROR OF COMMISSION in HYPOTHESIS TESTING.

β beta, lowercase b. Shorthand for ERROR OF OMISSION in hypothesis testing.

χ chi, lowercase c. See CHI-SQUARED TEST.

λ lambda, lowercase l. See QUEUEING.

μ mu, lowercase m. Shorthand for MEAN of a POPULATION. Also used in QUEUEING.

π pi, lowercase p. The constant 3.1415927... which is the ratio between the circumference and the radius of a circle (see AREA AND VOLUME). In statistics, π is used as shorthand for a PROPORTION of a population.

σ sigma, lowercase s. Shorthand for STANDARD DEVIATION of a population.

Σ Sigma, uppercase S. Shorthand for take the sum of.

T

t-ratio. A t-TEST score produced during REGRESSION ANALYSIS. As a rule of thumb, discard any INDEPENDENT VARIABLE that has a t-ratio between −2 and +2. Such a variable does not have a significant effect in explaining the DEPENDENT VARIABLE at a 95% level of CONFIDENCE. Include variables with t-ratios below −2 or above 2. Common sense may override this rule of thumb.

t-test. A STATISTICAL TEST used to analyse the MEAN of small SAMPLES. It approximates to the simple Z SCORE TEST of the NORMAL DISTRIBUTION once the sample is larger than about 30 items, which is a good reason to take samples of at least this size.

Table. An ordered presentation of related information. Chapter 4 indicates methods of interpreting and compiling tables. Tables of statistical information and other data (such as financial RATES OF RETURN) are available off the shelf. A regular table is also known as a MATRIX.

Tight. See SLACK.

Time series. A record of developments over time (eg, inflation, monthly output of widgets). Time series may record volumes (number of items produced), values (retail income from selling items) or prices. Volume times price equals value. Note that a time series may be split into a TREND, a CYCLE, a seasonal and a RESIDUAL (random, unidentifiable) component (see DECOMPOSITION, SEASONALITY). Compare with CROSS-SECTIONAL DATA (eg, output by production process).

Transfer pricing. The price charged by one part of a company to another on the transfer of goods of services. This can shift profits/losses around in an arbitrary manner for tax or other purposes.

Transformation. A bit of numerical trickery applied to a set of numbers to make their interpretation and analysis easier. For example, a series which is growing steadily will produce a curved line on a GRAPH. Convert this into a straight line with a log transformation by plotting LOGARITHMS of the numbers rather than the numbers themselves (see Figure 5.10). As with many numerical methods, you should not rely on a transformation unless you can come up with an appropriate rationale for picking it.

Trend. The long-run path of a TIME SERIES, which might be concealed by cyclical, seasonal and RANDOM or RESIDUAL variation (see also CYCLE, SEASONALITY). Use simple MOVING AVERAGES or EXPONENTIAL SMOOTHING, or more complex DECOMPOSITION to identify a trend. If projecting it forward, remember that it might be undergoing a change in direction.

Triangles and trigonometry. Some old hulks carry a huge tub of ale is about all you need to know to make occasional use of the sin, cos and tan keys on your calculator. The first letter of each word spells out the schoolbook formulae for dealing with triangles which have a right angle (90°) at one corner.

$\sin X° = O \div H$
$\cos X° = A \div H$
$\tan X° = O \div A$

H (= length of hypotenuse)

O (= length of opposite side)

Angle X°

90°

A (= length of adjacent side)

If you know one angle (other than the right angle) and one length, or two lengths, you can use these formulae to work out all the other angles and lengths. This is sometimes useful for obscure business problems involving measurements (see also AREA AND VOLUME). If a triangle does not have a right angle, divide it into two triangles that do. One other useful relationship for any triangle with angles A°, B° and C° and sides of lengths a, b, and c is:

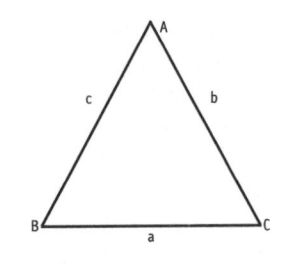

$$a \div \sin A° = b \div \sin B° = c \div \sin C°$$

Again, if any two are known, all the rest can be calculated.

Two-tail test. A statistical decision-making process which analyses variation in either direction from a point estimate. A packer guaranteeing average contents for bags of sweets is interested in both tails of a DISTRIBUTION since a pack with too many or too few items will be costly one way or another. Compare with a ONE-TAIL TEST. See SAMPLING.

Type I and II errors. See ERROR OF COMMISSION and ERROR OF OMISSION.

U

Unbounded. An infinitely large set of feasible solutions to a mathematical programming problem; it may be impossible to reach the optimal solution. But check that an apparently unbounded problem does not reflect a keying or computational error.

Uncertainty. Unquantified RISK. DECISION ANALYSIS provides a framework for decision-making under uncertainty, but there are much better techniques for decision-making when uncertainty is quantified (ie, as risk) using PROBABILITIES.

Unconditional probability. The PROBABILITY of an event when it is not dependent on another event occurring also. The opposite is conditional probability.

Uncontrollables. See SITUATION.

Unit normal loss. Loss per unit when sales follow a NORMAL DISTRI-BUTION and are below the BREAK EVEN point. It reveals the maximum amount to spend on obtaining extra information (the EXPECTED VALUE OF PERFECT INFORMATION).

Utility. A subjective and personal measure of underlying value. Cash-poor companies attach a much higher utility to gaining an extra $1 than do prosperous ones. Similarly, a loss of $10m might be more than ten times as serious as a loss of $1m. Table 7.5 shows how to assess utility and the associated text shows how to incorporate it in the decision-making process.

V

Value and volume. Volume records quantity, say the number of items sold in a given period. Value is the volume multiplied by the price. In general, analyse volumes and values separately, then the distinction between the effects of changes in price and changes in volume is clear. Similarly, prefer to forecast in volume terms, then adjust for INFLA-TION. Values are also known as CURRENT PRICES and nominal prices. When volumes are measured in money terms (in prices ruling on one particular date, perhaps a BASE year), they are said to be in CONSTANT PRICES or in real terms.

Value of a game. Think of this as the PAYOFF which changes hands in a competitive game (see GAME STRATEGY). In a STRICTLY DETER-MINED game the value is identified by the SADDLE POINT.

Variable. Anything that varies (such as profits and sales) as opposed to items which are CONSTANT (eg, price per unit), for the purposes of a particular piece of analysis.

Variable costs. Costs which are directly related to the cost of produc-tion (such as raw materials). Compare with FIXED COSTS. See ABSORP-TION COSTING and MARGINAL ANALYSIS.

Variance (1). Difference between BUDGET and OUTTURN. Variances highlight deviations from plan, allowing corrective action to be taken.

Variance (II). A measure of the SPREAD of a set of numbers quoted in squared units. For example, the variance of a set of heights measured in feet is given in square feet. As this is hard to relate to, it is better to use the STANDARD DEVIATION (which is the square root of the variance).

Venn diagram. A pictorial way of coping with PROBABILITIES. It is easy to describe with a playing card problem. In the following venn diagram, the outer square represents all 52 playing cards. The left circle is the set of 13 hearts, and the right circle is the set of 4 kings. The overlap is obviously the king of hearts, which is in both sets (circles). The diagram helps identify, for example, that the probability of drawing a heart or a king is $(13 \div 52) + (4 \div 52) - (1 \div 52) = 16 \div 52$. The $(1 \div 52)$ is subtracted because otherwise the king of hearts is double counted. For more complex problems, drawing venn diagrams and writing in quantities or probabilities helps to reach the correct solution.

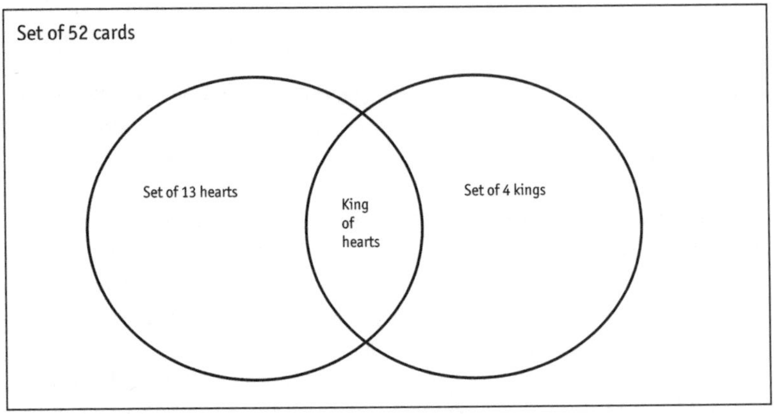

Set of 52 cards

Set of 13 hearts

King of hearts

Set of 4 kings

Volume. See VALUE AND VOLUME or AREA AND VOLUME.

W

Waiting lines. See QUEUEING.

Weighted average. An AVERAGE in which the components are scaled to reflect their relative importance. If three packets of peanuts (profit €1.50 per pack) are sold for each packet of cashew nuts (profit €1.25 per pack), the weighted average profit per pack from the two flavours is $(€1.50 \times 0.75) + (€1.25 \times 0.25) = €1.44$. Making the weights sum to one simplifies the arithmetic. Stockmarket and exchange rate indices are nearly all weighted averages.

Written down value. See NET BOOK VALUE.

X

x and y. Letters such as x and y near the end of the alphabet are often used to identify VARIABLES. It is much easier to note y = ... rather than keep writing profits = ... or costs = ..., etc. In REGRESSION ANALYSIS, y is frequently used to denote the DEPENDENT VARIABLE (eg, profits) while x denotes the INDEPENDENT VARIABLE (eg, units sold).

Y

Yield. Returns from an investment measured in PERCENTAGE terms. Compare using EFFECTIVE INTEREST RATES, analyse with INVEST-MENT ANALYSIS techniques.

Z

Zero sum game. A game where one player's loss is matched exactly by another's gain. For example, if company A gains 10% of the market share, company B loses that amount. See GAME STRATEGY.

z score (I). A standardised STANDARD DEVIATION, used to locate a point in the NORMAL DISTRIBUTION. The relationship is $z = (x - \mu) \div \sigma$ where σ is the standard deviation, μ is the MEAN, and x is the point to be located. For example, if sales are expected to average 10,000 units with a standard deviation of 1,500 units, then for x = 7,750 units, $z = (7,750 - 10,000) \div 1,500 = -1.50$. Normal tables (Table 3.2), also known as z tables, will reveal the PROPORTION of the distribution on either side of this point (6.7% and 93.3%). This might tell the business managers that they have a 6.7% RISK of failing to break even. Quite apart from the importance of z scores for analysing normal distributions (including those used to model risk), z is also used in STATISTICAL TESTS of SAMPLE means.

z score (II). A WEIGHTED AVERAGE of the factors which go into DIS-CRIMINANT ANALYSIS. z scores were once popular, and could be again, for combining RATIOS (such as the CURRENT RATIO and PROFIT MARGIN) to obtain a measure of the likelihood of a company going bust.

Index